1988
YEARBOOK of ASTRONOMY

1988 YEARBOOK of ASTRONOMY

edited by
Patrick Moore

W·W·Norton & Company
NEW YORK LONDON

Copyright © 1987 by Sidgwick and Jackson Limited

First American edition, 1988

All rights reserved

ISBN 0-393-02526-8

W. W. Norton & Company, Inc.,
500 Fifth Avenue, New York, NY 10110
W. W. Norton & Company, Ltd.,
37 Great Russell Street, London WC1B 3NU

Printed in Great Britain
1 2 3 4 5 6 7 8 9 0

Contents

Editor's Foreword . 7

Preface . 9

Part One: *Monthly Charts and Astronomical Phenomena*

Notes on the Star Charts . 11

Northern Star Charts . 14

Southern Star Charts . 40

The Planets and the Ecliptic . 66

Phases of the Moon, 1988 . 71

Longitudes of the Sun, Moon, and Planets in 1988 . 72

Events in 1988 . 74

Monthly Notes . 75

Eclipses in 1988 . 115

Occultations in 1988 . 117

Comets in 1988 . 119

Minor Planets in 1988 . 120

Meteors in 1988 . 121

Some Events in 1989 . 122

Part Two: *Article Section*

The William Herschel Telescope
  Paul Murdin and *Alec Boksenberg* . 125

The Shadow Chasers
  David A. Allen . 135

The Future Exploration of Mars
  Garry E. Hunt . 147

Background to the Big Bang Theory
  Ron C. Maddison . 156

A Glow from the Past
  Alan E. Wright . 172

Supernovæ
  Patrick Moore . 176

Part Three: *Miscellaneous*

Some Interesting Telescopic Variable Stars . 183

Some Interesting Double Stars . 185

Some Interesting Nebulæ and Clusters . 187

Our Contributors . 189

Editor's Foreword

The twenty-sixth *Yearbook* follows the usual pattern; there seems no reason to alter it. Much has happened during the past year, and in our article section we have concentrated upon the new developments – notably the William Herschel Telescope, which is just embarking upon what will certainly be a glorious career, and which is described in this issue by the Director of the Royal Greenwich Observatory and the Project Scientist who has been in charge of our telescopes at La Palma in the Canary Islands. Mars, too, is very much in the news; 1988 sees the most favourable opposition for many years, and we welcome an article from another of our regular contributors, Dr Garry Hunt, one of NASA's Principal Scientific Investigators. We again welcome Dr David Allen and Dr Alan Wright, from Australia, and that well-known lecturer and broadcaster Dr Ron Maddison of Keele.

This issue goes to press in the summer of 1987. There is no knowing what will happen before publication day, but we keep as up to date as possible; and 'current events' are described in regular journals such as *Sky and Telescope* (USA) and *Astronomy Now* (the new periodical published in London).

As always, the monthly notes and charts have been provided by Gordon Taylor, without whom the task of producing this *Yearbook* would indeed have been difficult. Unfortunately his name was omitted from the Acknowledgements in the *1986 Yearbook* – an omission for which I was responsible; I simply committed the sin of 'overlooking the obvious', though I am glad to say that I have been forgiven!

PATRICK MOORE

Selsey, April 1987

Preface

New readers will find that all the information in this *Yearbook* is given in diagrammatic or descriptive form; the positions of the planets may easily be found on the specially designed star charts, while the monthly notes describe the movements of the planets and give details of other astronomical phenomena visible in both the northern and southern hemispheres. Two sets of the star charts are provided. The **Northern Charts** (pp. 14 to 39) are designed for use in latitude 52 degrees north, but may be used without alteration throughout the British Isles, and (except in the case of eclipses and occultations) in other countries of similar north latitude. The **Southern Charts** (pp. 40 to 65) are drawn for latitude 35 degrees south, and are suitable for use in South Africa, Australia and New Zealand, and other stations in approximately the same south latitude. The reader who needs more detailed information will find *Norton's Star Atlas* (Longman) an invaluable guide, while more precise positions of the planets and their satellites, together with predictions of occultations, meteor showers, and periodic comets may be found in the *Handbook* of the British Astronomical Association. A somewhat similar publication is the *Observer's Handbook* of the Royal Astronomical Society of Canada, and readers will also find details of forthcoming events given in the American *Sky and Telescope*. This monthly publication also produces a special occultation supplement giving predictions for the United States and Canada. The British equivalent is the new periodical *Astronomy Now*.

Important Note
The times given on the star charts and in the Monthly Notes are generally given as local times, using the 24-hour clock, the day beginning at midnight. All the dates, and the times of a few events

(e.g. eclipses), are given in Greenwich Mean Time (G.M.T.), which is related to local time by the formula

Local Mean Time = G.M.T. − west longitude

In practice, small differences of longitudes are ignored, and the observer will use local clock time, which will be the appropriate Standard (or Zone) Time. As the formula indicates, places in west longitude will have a Standard Time slow on G.M.T., while places in east longitude will have a Standard Time fast on G.M.T. as examples we have:

Standard Time in

New Zealand	G.M.T.	+	12 hours
Victoria; N.S.W.	G.M.T.	+	10 hours
Western Australia	G.M.T.	+	8 hours
South Africa	G.M.T.	+	2 hours
British Isles	G.M.T.		
Eastern S.T.	G.M.T.	−	5 hours
Central S.T.	G.M.T.	−	6 hours, etc

If Summer Time is in use, the clocks will have to have been advanced by one hour, and this hour must be subtracted from the clock time to give Standard Time.

In Great Britain and N. Ireland, Summer Time will be in force in 1988 from March $27^{d}01^{h}$ until October $23^{d}01^{h}$ G.M.T.

PART ONE

Monthly Charts and Astronomical Phenomena

Notes on the Star Charts

The stars, together with the Sun, Moon and planets seem to be set on the surface of the celestial sphere, which appears to rotate about the Earth from east to west. Since it is impossible to represent a curved surface accurately on a plane, any kind of star map is bound to contain some form of distortion. But it is well known that the eye can endure some kinds of distortion better than others, and it is particularly true that the eye is most sensitive to deviations from the vertical and horizontal. For this reason the star charts given in this volume have been designed to give a true representation of vertical and horizontal lines, whatever may be the resulting distortion in the shape of a constellation figure. It will be found that the amount of distortion is, in general, quite small, and is only obvious in the case of large constellations such as Leo and Pegasus, when these appear at the top of the charts, and so are drawn out sideways.

The charts show all stars down to the fourth magnitude, together with a number of fainter stars which are necessary to define the shape of a constellation. There is no standard system for representing the outlines of the constellations, and triangles and other simple figures have been used to give outlines which are easy to follow with the naked eye. The names of the constellations are given, together with the proper names of the brighter stars. The apparent magnitudes of the stars are indicated roughly by using four different sizes of dots, the larger dots representing the bright stars.

The two sets of star charts are similar in design. At each opening there is a group of four charts which give a complete coverage of the sky up to an altitude of 62½ degrees; there are twelve such groups to cover the entire year. In the **Northern Charts** (for 52 degrees north) the upper two charts show the southern sky, south being at the centre and east on the left. The coverage is from 10 degrees north of east (top left) to 10 degrees north of west (top right). The two lower charts show the northern sky from 10 degrees south of west (lower left) to 10 degrees south of east (lower right). There is thus an overlap east and west.

Conversely, in the **Southern Charts** (for 35 degrees south) the upper two charts show the northern sky, with north at the centre and east on the right. The two lower charts show the southern sky, with south at the centre and east on the left. The coverage and overlap is the same on both sets of charts.

Because the sidereal day is shorter than the solar day, the stars appear to rise and set about four minutes earlier each day, and this amounts to two hours in a month. Hence the twelve groups of charts in each set are sufficient to give the appearance of the sky throughout the day at intervals of two hours, or at the same time of night at monthly intervals throughout the year. The actual range of dates and times when the stars on the charts are visible is indicated at the top of each page. Each group is numbered in bold type, and the number to be used for any given month and time is summarized in the following table:

Local Time	18h	20h	22h	0h	2h	4h	6h
January	11	12	1	2	3	4	5
February	12	1	2	3	4	5	6
March	1	2	3	4	5	6	7
April	2	3	4	5	6	7	8
May	3	4	5	6	7	8	9
June	4	5	6	7	8	9	10
July	5	6	7	8	9	10	11
August	6	7	8	9	10	11	12
September	7	8	9	10	11	12	1
October	8	9	10	11	12	1	2
November	9	10	11	12	1	2	3
December	10	11	12	1	2	3	4

The charts are drawn to scale, the horizontal measurements, marked at every 10 degrees, giving the azimuths (or true bearings) measured from the north round through east (90 degrees), south (180 degrees), and west (270 degrees). The vertical measurements, similarly marked, give the altitudes of the stars up to 62½ degrees. Estimates of altitude and azimuth made from these charts will necessarily be mere approximations, since no observer will be exactly at the adopted latitude, or at the stated time, but they will serve for the identification of stars and planets.

The ecliptic is drawn as a broken line on which longitude is marked at every 10 degrees; the positions of the planets are then easily found by reference to the table on page 72. It will be noticed that on the Southern Charts the **ecliptic** may reach an altitude in excess of 62½ degrees on star charts 5 to 9. The continuations of the broken line will be found on the charts of overhead stars.

There is a curious illusion that stars at an altitude of 60 degrees or more are actually overhead, and the beginner may often feel that he is leaning over backwards in trying to see them. These overhead stars are given separately on the pages immediately following the main star charts. The entire year is covered at one opening, each of the four maps showing the overhead stars at times which correspond to those of three of the main star charts. The position of the zenith is indicated by a cross, and this cross marks the centre of a circle which is 35 degrees from the zenith; there is thus a small overlap with the main charts.

The broken line leading from the north (on the Northern Charts) or from the south (on the Southern Charts) is numbered to indicate the corresponding main chart. Thus on page 38 the N-S line numbered 6 is to be regarded as an extension of the centre (south) line of chart 6 on pages 24 and 25, and at the top of these pages are printed the dates and times which are appropriate. Similarly, on page 65, the S-N line numbered 10 connects with the north line of the upper charts on pages 58 and 59.

The overhead stars are plotted as maps on a conical projection, and the scale is rather smaller than that of the main charts.

1L

October 6 at 5^h	October 21 at 4^h
November 6 at 3^h	November 21 at 2^h
December 6 at 1^h	December 21 at midnight
January 6 at 23^h	January 21 at 22^h
February 6 at 21^h	February 21 at 20^h

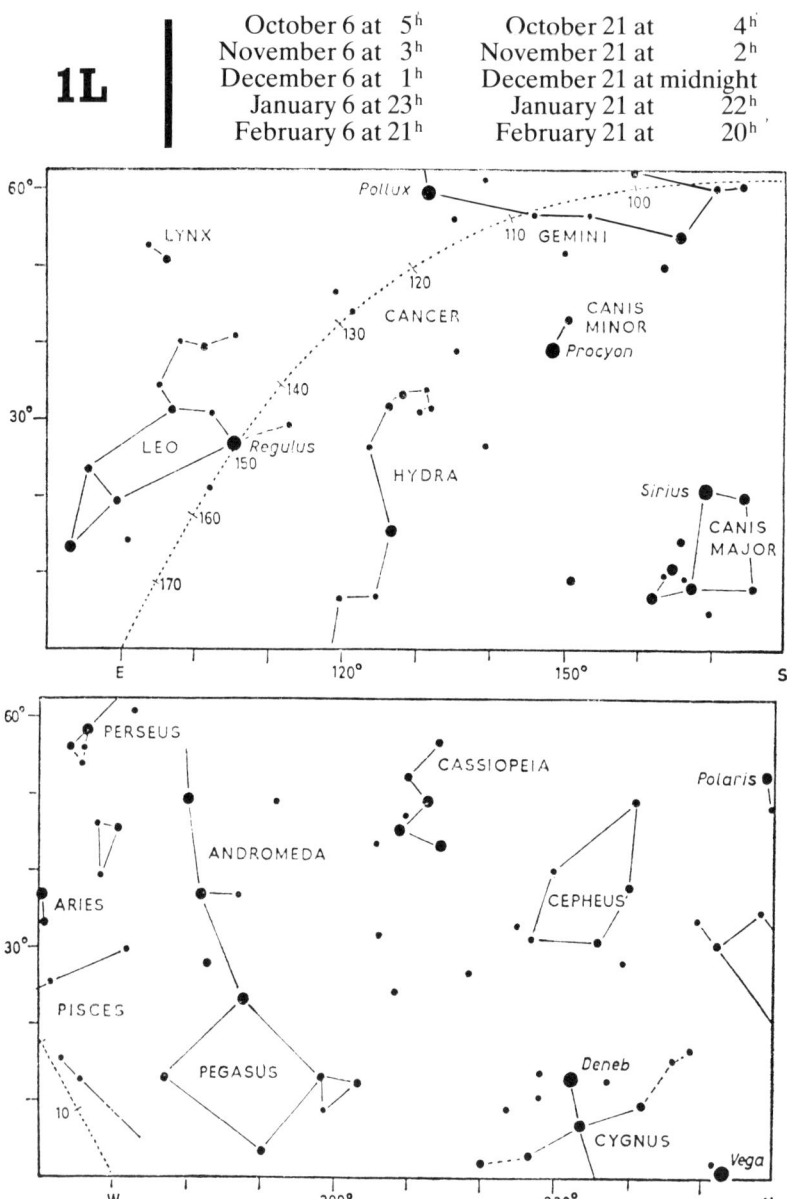

NORTHERN STAR CHARTS

October 6 at 5ʰ   October 21 at 4ʰ
November 6 at 3ʰ   November 21 at 2ʰ
December 6 at 1ʰ   December 21 at midnight
January 6 at 23ʰ   January 21 at 22ʰ
February 6 at 21ʰ February 21 at 20ʰ

1R

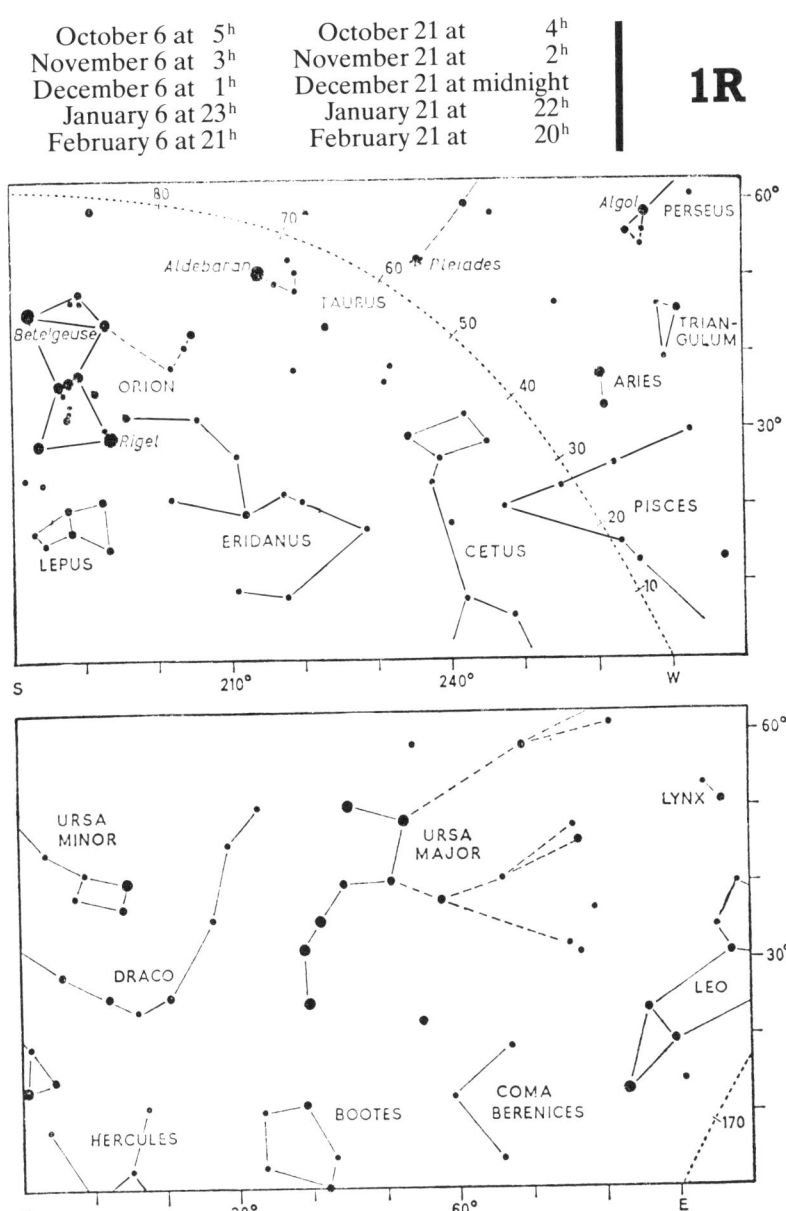

2L

November 6 at 5^h	November 21 at 4^h
December 6 at 3^h	December 21 at 2^h
January 6 at 1^h	January 21 at midnight
February 6 at 23^h	February 21 at 22^h
March 6 at 21^h	March 21 at 20^h

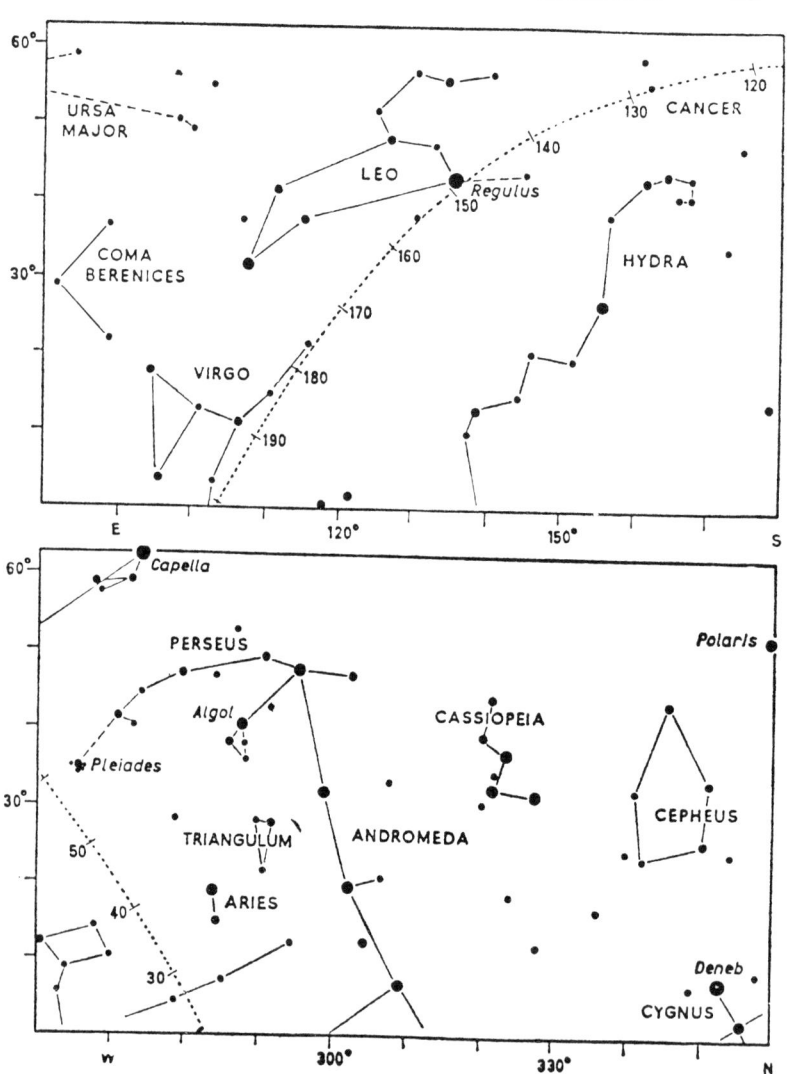

NORTHERN STAR CHARTS

November 6 at 5ʰ November 21 at 4ʰ
December 6 at 3ʰ December 21 at 2ʰ
January 6 at 1ʰ   January 21 at midnight
February 6 at 23ʰ February 21 at 22ʰ
March 6 at 21ʰ   March 21 at 20ʰ

2R

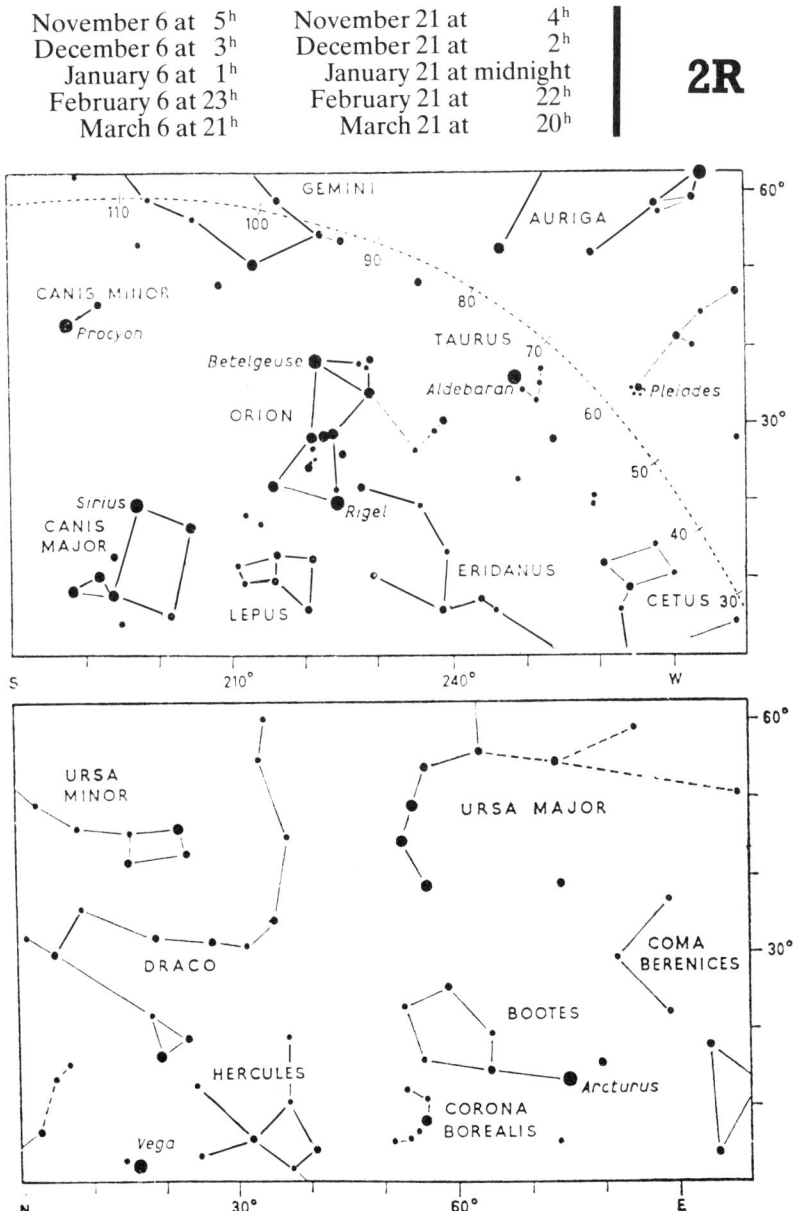

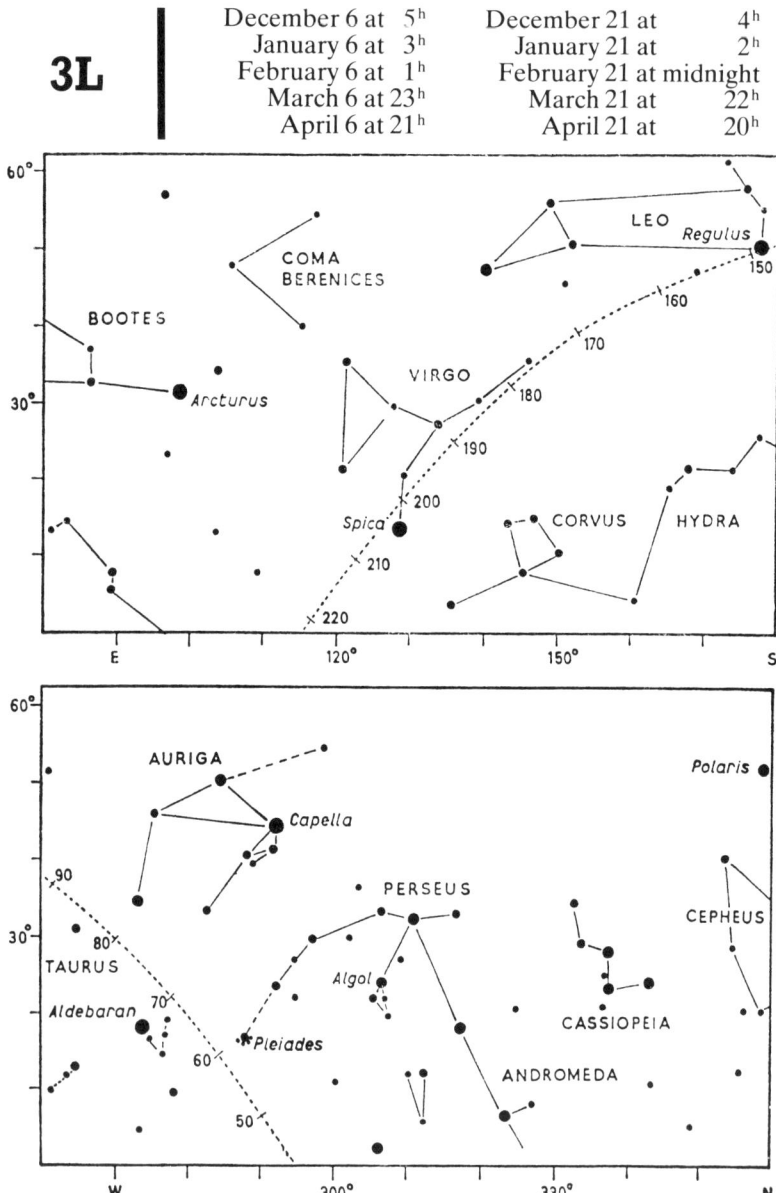

NORTHERN STAR CHARTS

December 6 at 5ʰ
January 6 at 3ʰ
February 6 at 1ʰ
March 6 at 23ʰ
April 6 at 21ʰ

December 21 at 4ʰ
January 21 at 2ʰ
February 21 at midnight
March 21 at 22ʰ
April 21 at 20ʰ

3R

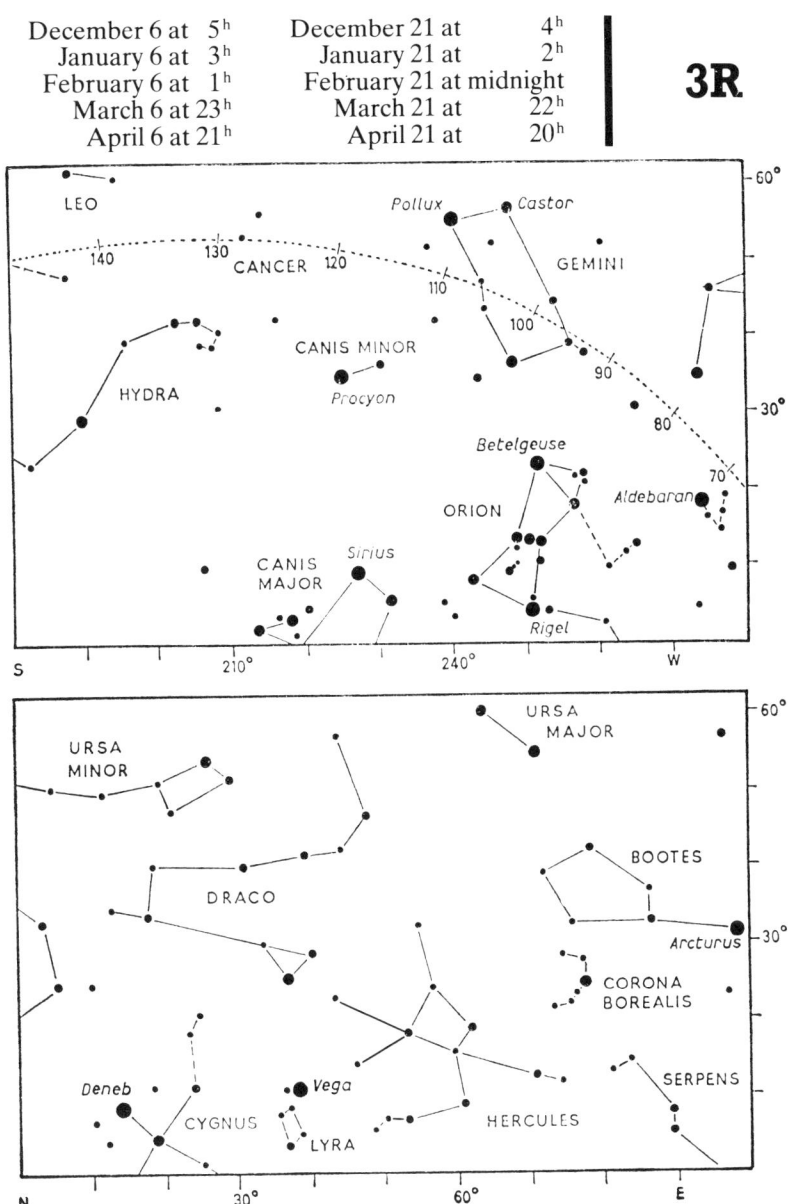

4L

January 6 at 5ʰ	January 21 at 4ʰ
February 6 at 3ʰ	February 21 at 2ʰ
March 6 at 1ʰ	March 21 at midnight
April 6 at 23ʰ	April 21 at 22ʰ
May 6 at 21ʰ	May 21 at 20ʰ

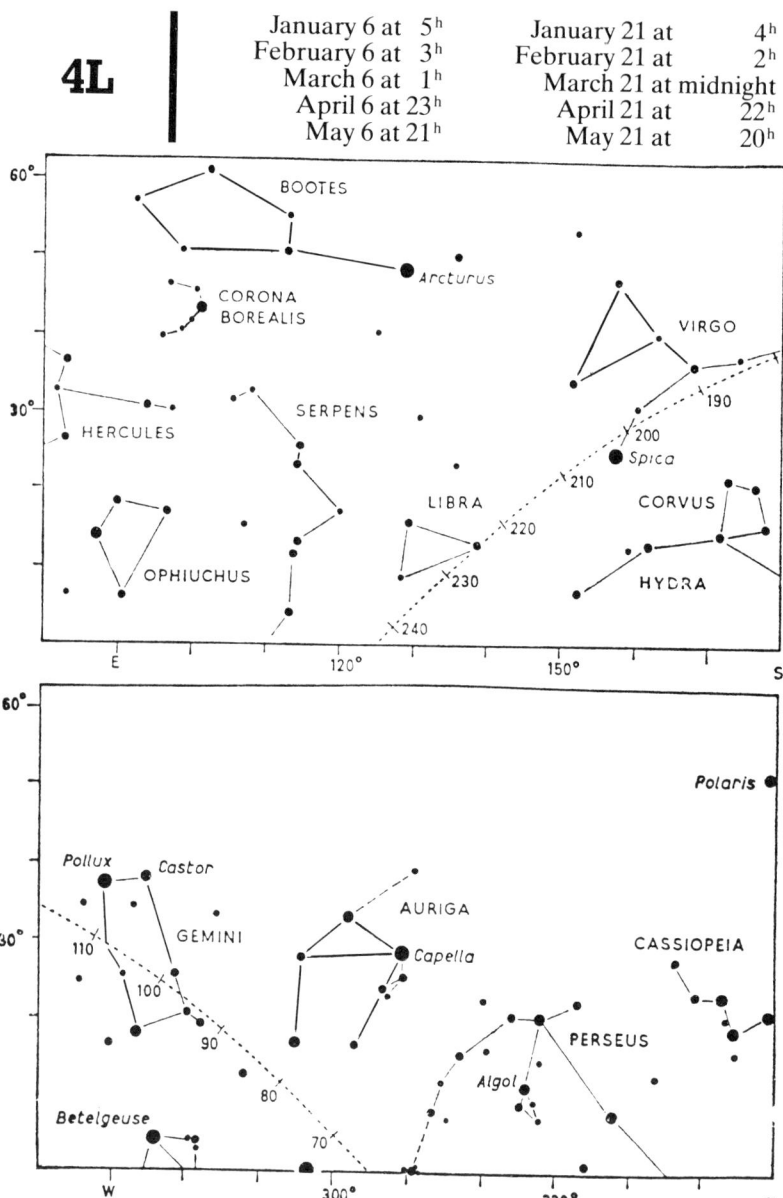

NORTHERN STAR CHARTS

January 6 at 5ʰ
February 6 at 3ʰ
March 6 at 1ʰ
April 6 at 23ʰ
May 6 at 21ʰ

January 21 at 4ʰ
February 21 at 2ʰ
March 21 at midnight
April 21 at 22ʰ
May 21 at 20ʰ

4R

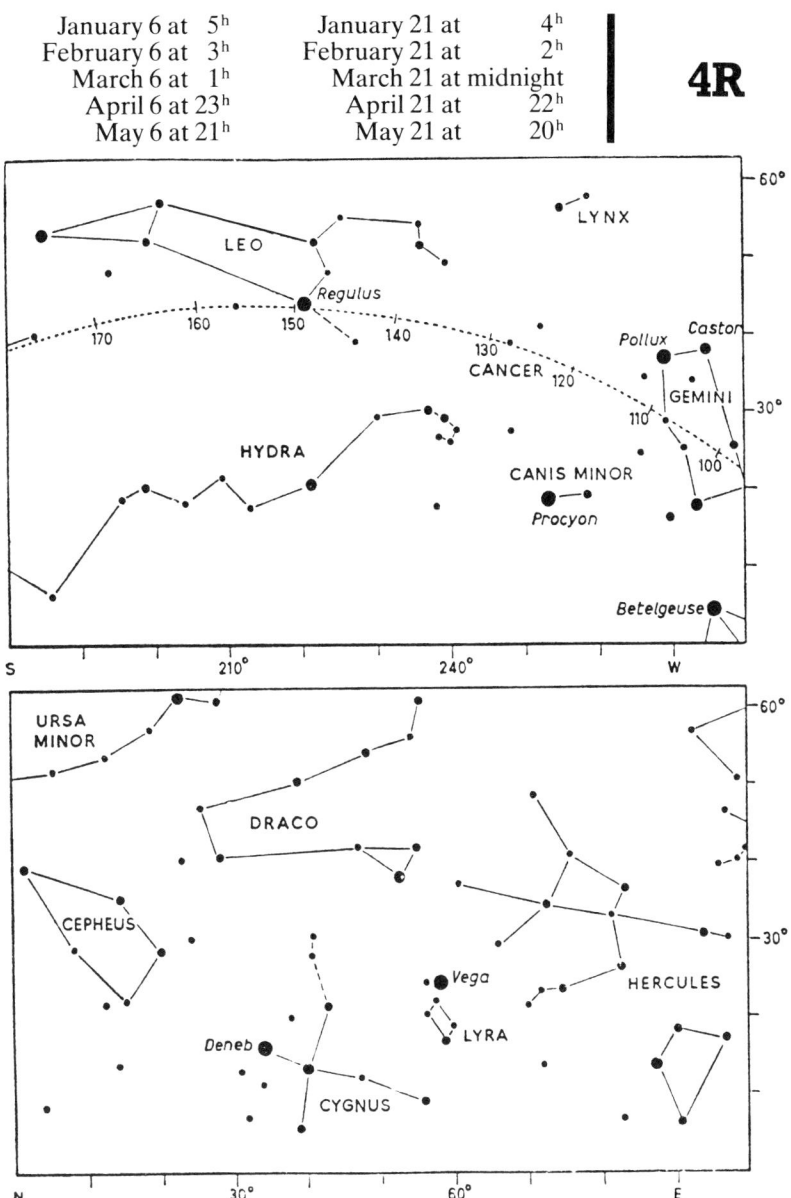

1988 YEARBOOK OF ASTRONOMY

5L

January 6 at 7ʰ	January 21 at 6ʰ
February 6 at 5ʰ	February 21 at 4ʰ
March 6 at 3ʰ	March 21 at 2ʰ
April 6 at 1ʰ	April 21 at midnight
May 6 at 23ʰ	May 21 at 22ʰ

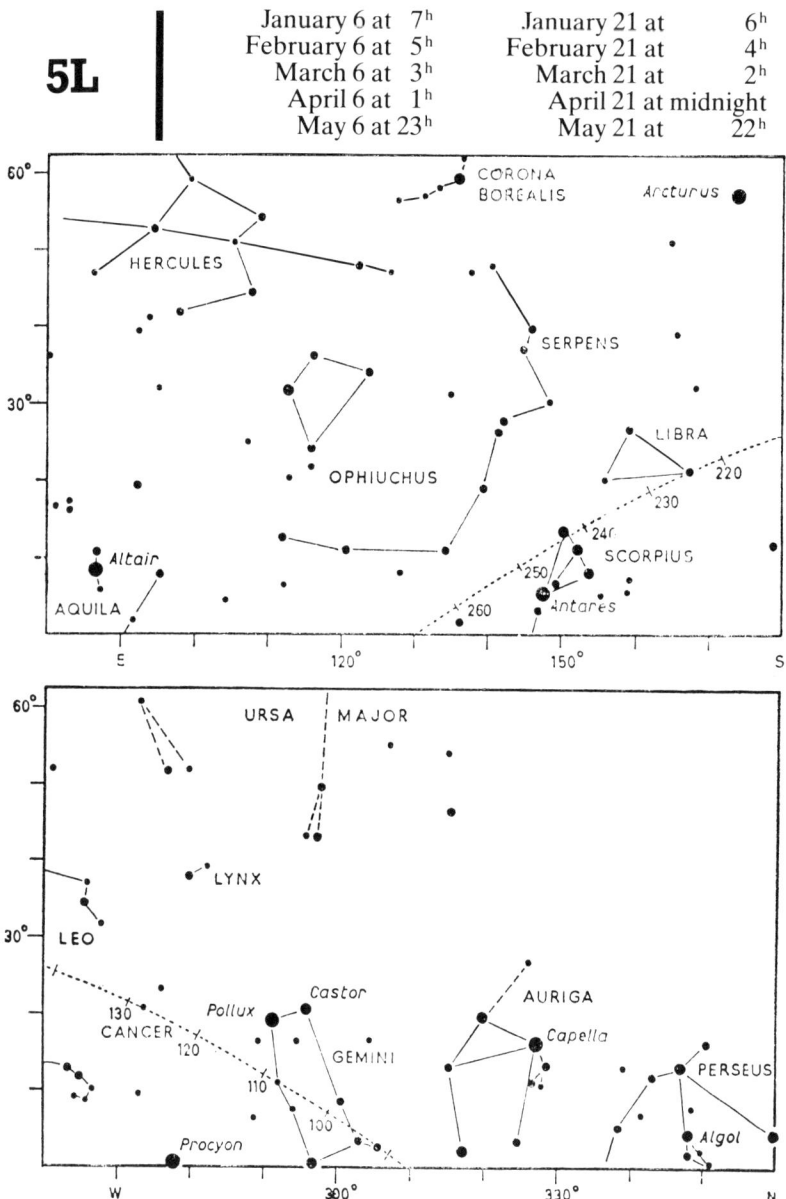

NORTHERN STAR CHARTS

January 6 at 7ʰ   January 21 at 6ʰ
February 6 at 5ʰ February 21 at 4ʰ
March 6 at 3ʰ   March 21 at 2ʰ
April 6 at 1ʰ   April 21 at midnight
May 6 at 23ʰ   May 21 at 22ʰ

5R

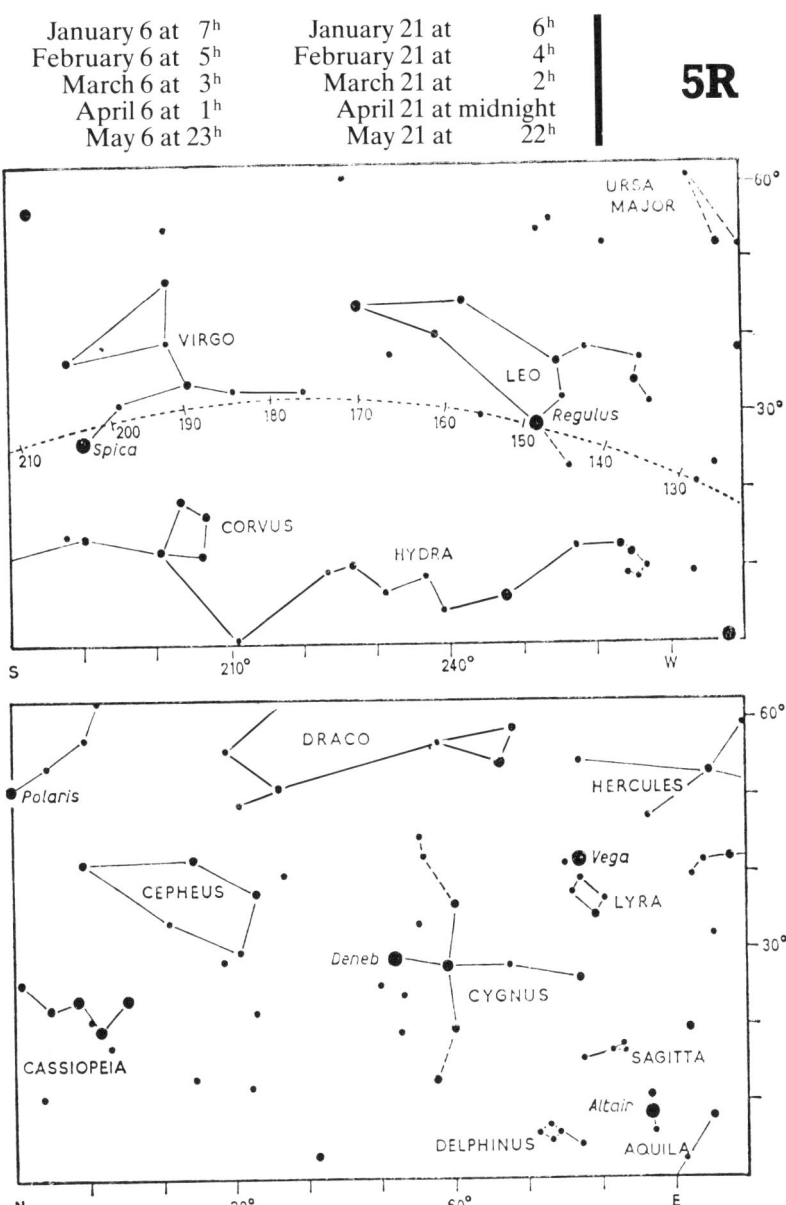

1988 YEARBOOK OF ASTRONOMY

6L

March 6 at 5ʰ	March 21 at 4ʰ
April 6 at 3ʰ	April 21 at 2ʰ
May 6 at 1ʰ	May 21 at midnight
June 6 at 23ʰ	June 21 at 22ʰ
July 6 at 21ʰ	July 21 at 20ʰ

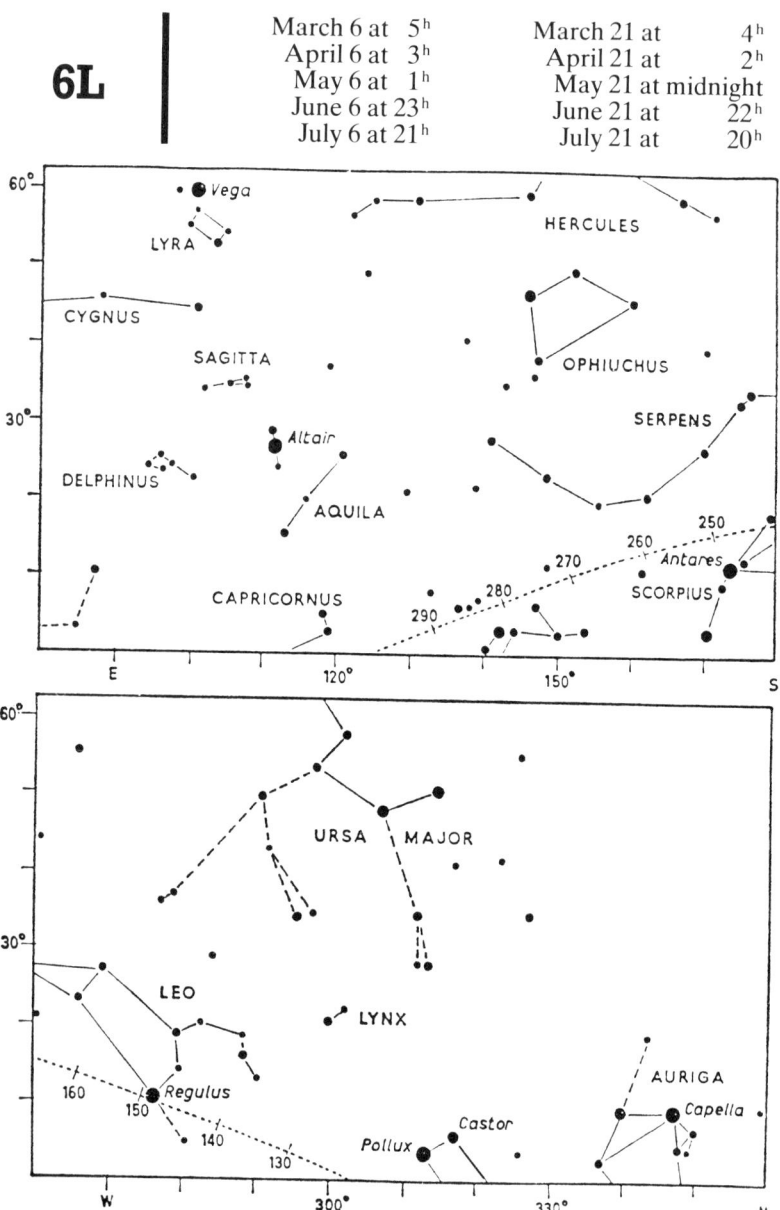

NORTHERN STAR CHARTS

March 6 at 5ʰ March 21 at 4ʰ
April 6 at 3ʰ April 21 at 2ʰ
May 6 at 1ʰ   May 21 at midnight
June 6 at 23ʰ June 21 at 22ʰ
July 6 at 21ʰ July 21 at 20ʰ

6R

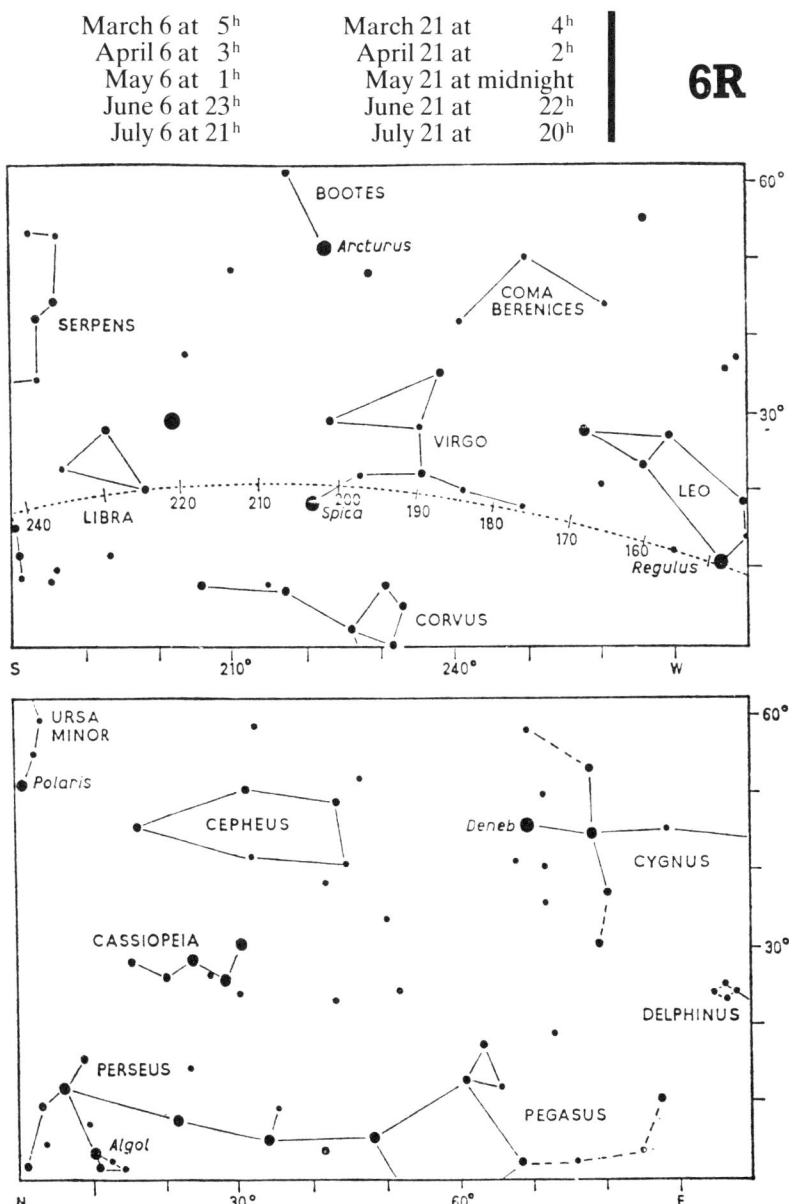

7L

May 6 at 3ʰ	May 21 at 2ʰ
June 6 at 1ʰ	June 21 at midnight
July 6 at 23ʰ	July 21 at 22ʰ
August 6 at 21ʰ	August 21 at 20ʰ
September 6 at 19ʰ	September 21 at 18ʰ

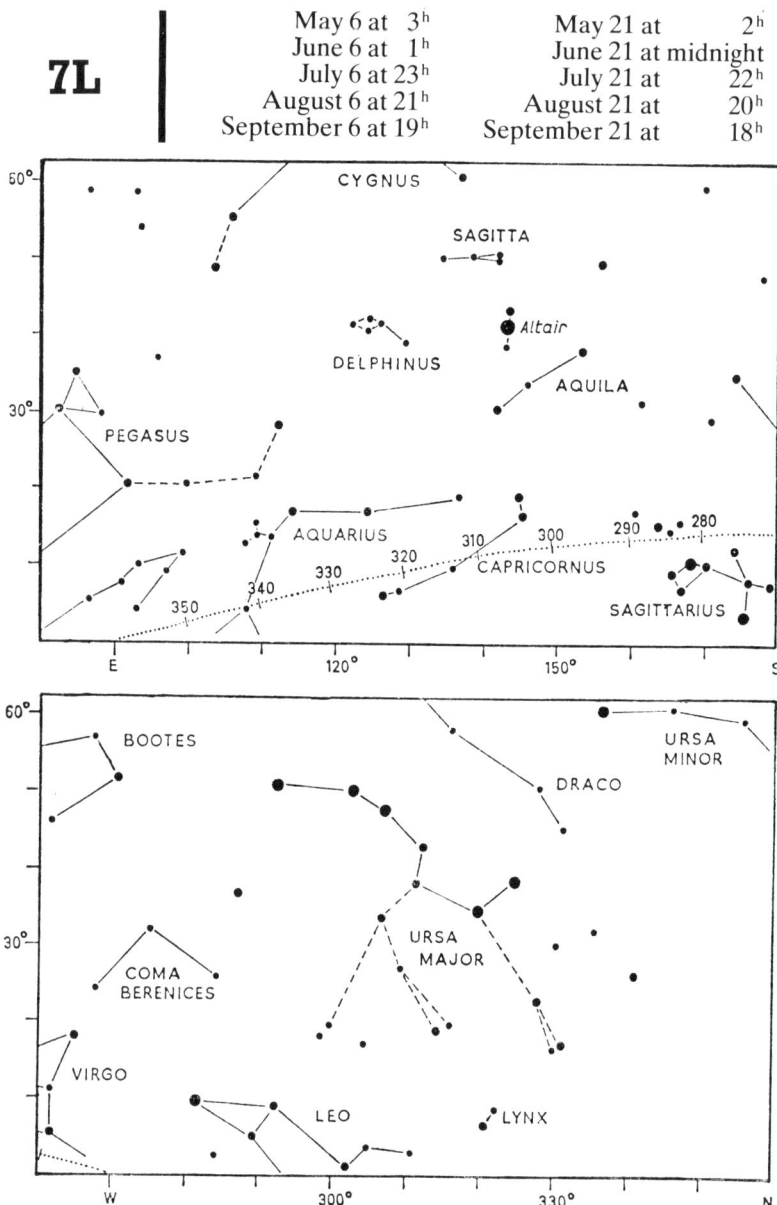

NORTHERN STAR CHARTS

May 6 at 3ʰ	May 21 at 2ʰ
June 6 at 1ʰ	June 21 at midnight
July 6 at 23ʰ	July 21 at 22ʰ
August 6 at 21ʰ	August 21 at 20ʰ
September 6 at 19ʰ	September 21 at 18ʰ

7R

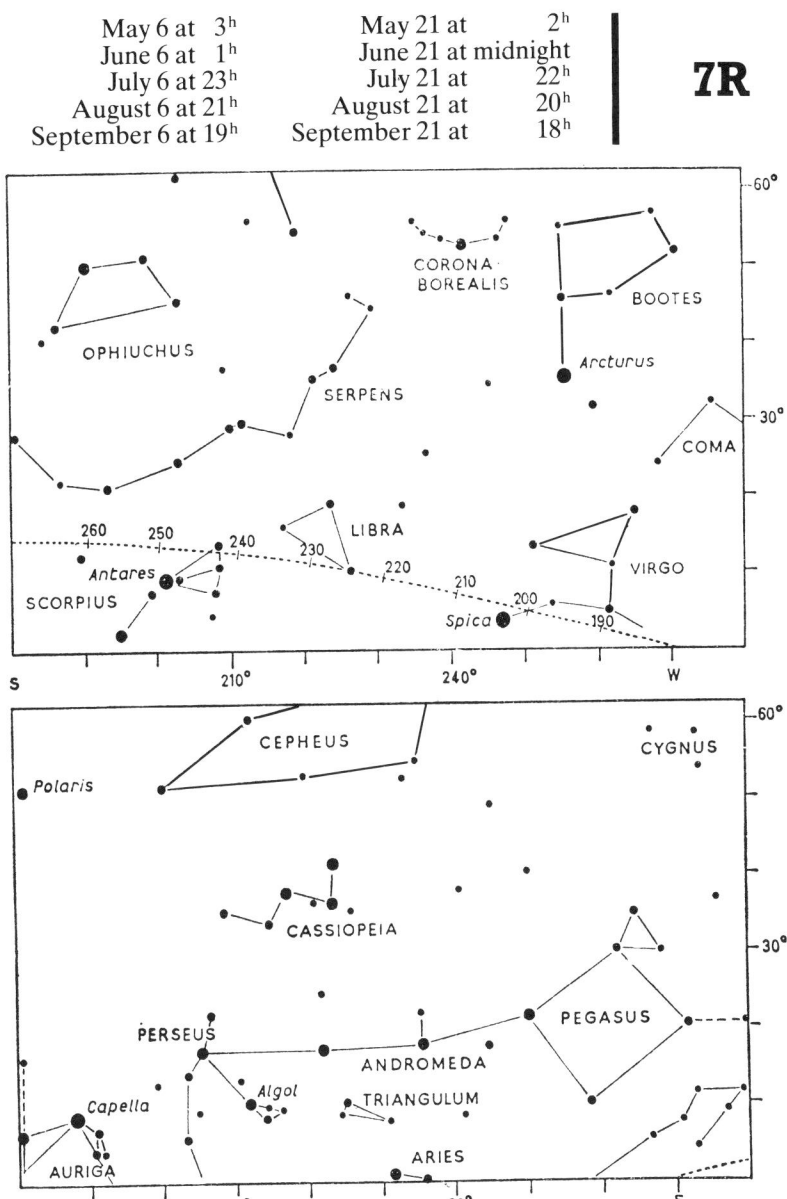

8L

July 6 at 1ʰ	July 21 at midnight
August 6 at 23ʰ	August 21 at 22ʰ
September 6 at 21ʰ	September 21 at 20ʰ
October 6 at 19ʰ	October 21 at 18ʰ
November 6 at 17ʰ	November 21 at 16ʰ

NORTHERN STAR CHARTS

July 6 at 1ʰ
August 6 at 23ʰ
September 6 at 21ʰ
October 6 at 19ʰ
November 6 at 17ʰ

July 21 at midnight
August 21 at 22ʰ
September 21 at 20ʰ
October 21 at 18ʰ
November 21 at 16ʰ

8R

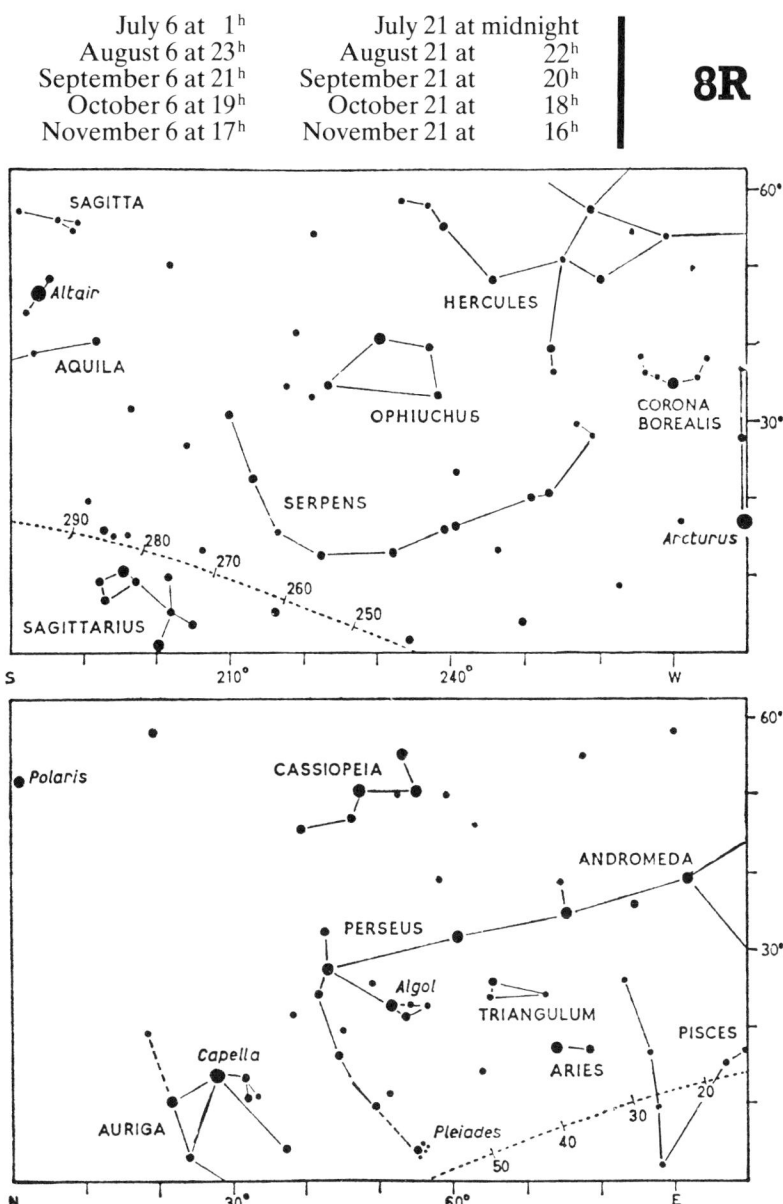

1988 YEARBOOK OF ASTRONOMY

9L

August 6 at 1ʰ	August 21 at midnight
September 6 at 23ʰ	September 21 at 22ʰ
October 6 at 21ʰ	October 21 at 20ʰ
November 6 at 19ʰ	November 21 at 18ʰ
December 6 at 17ʰ	December 21 at 16ʰ

NORTHERN STAR CHARTS

August 6 at 1ʰ	August 21 at midnight	
September 6 at 23ʰ	September 21 at 22ʰ	**9R**
October 6 at 21ʰ	October 21 at 20ʰ	
November 6 at 19ʰ	November 21 at 18ʰ	
December 6 at 17ʰ	December 21 at 16ʰ	

10L

August 6 at 3ʰ	August 21 at 2ʰ
September 6 at 1ʰ	September 21 at midnight
October 6 at 23ʰ	October 21 at 22ʰ
November 6 at 21ʰ	November 21 at 20ʰ
December 6 at 19ʰ	December 21 at 18ʰ

NORTHERN STAR CHARTS

August 6 at 3ʰ	August 21 at 2ʰ
September 6 at 1ʰ	September 21 at midnight
October 6 at 23ʰ	October 21 at 22ʰ
November 6 at 21ʰ	November 21 at 20ʰ
December 6 at 19ʰ	December 21 at 18ʰ

10R

11L

September 6 at 3^h	September 21 at 2^h
October 6 at 1^h	October 21 at midnight
November 6 at 23^h	November 21 at 22^h
December 6 at 21^h	December 21 at 20^h
January 6 at 19^h	January 21 at 18^h

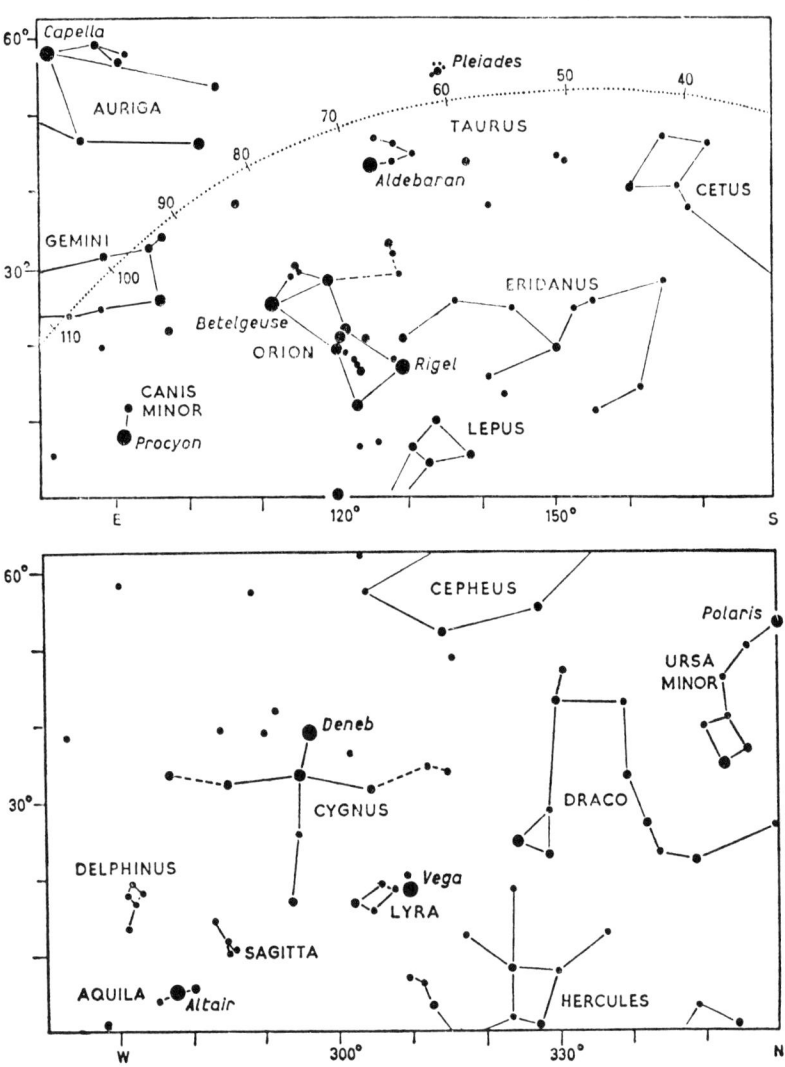

NORTHERN STAR CHARTS

September 6 at 3^h	September 21 at 2^h	
October 6 at 1^h	October 21 at midnight	
November 6 at 23^h	November 21 at 22^h	**11R**
December 6 at 21^h	December 21 at 20^h	
January 6 at 19^h	January 21 at 18^h	

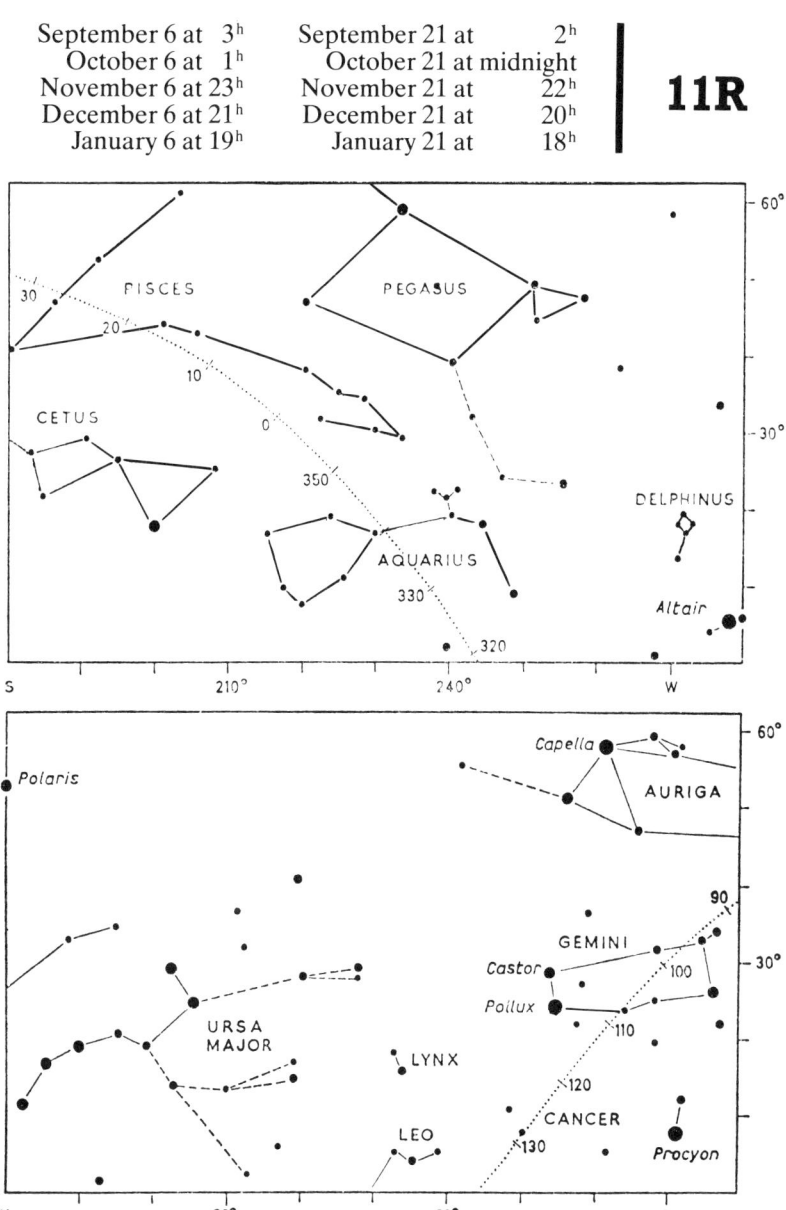

12L

October 6 at 3ʰ	October 21 at 2ʰ
November 6 at 1ʰ	November 21 at midnight
December 6 at 23ʰ	December 21 at 22ʰ
January 6 at 21ʰ	January 21 at 20ʰ
February 6 at 19ʰ	February 21 at 18ʰ

NORTHERN STAR CHARTS

October 6 at 3^h	October 21 at 2^h	
November 6 at 1^h	November 21 at midnight	**12R**
December 6 at 23^h	December 21 at 22^h	
January 6 at 21^h	January 21 at 20^h	
February 6 at 19^h	February 21 at 18^h	

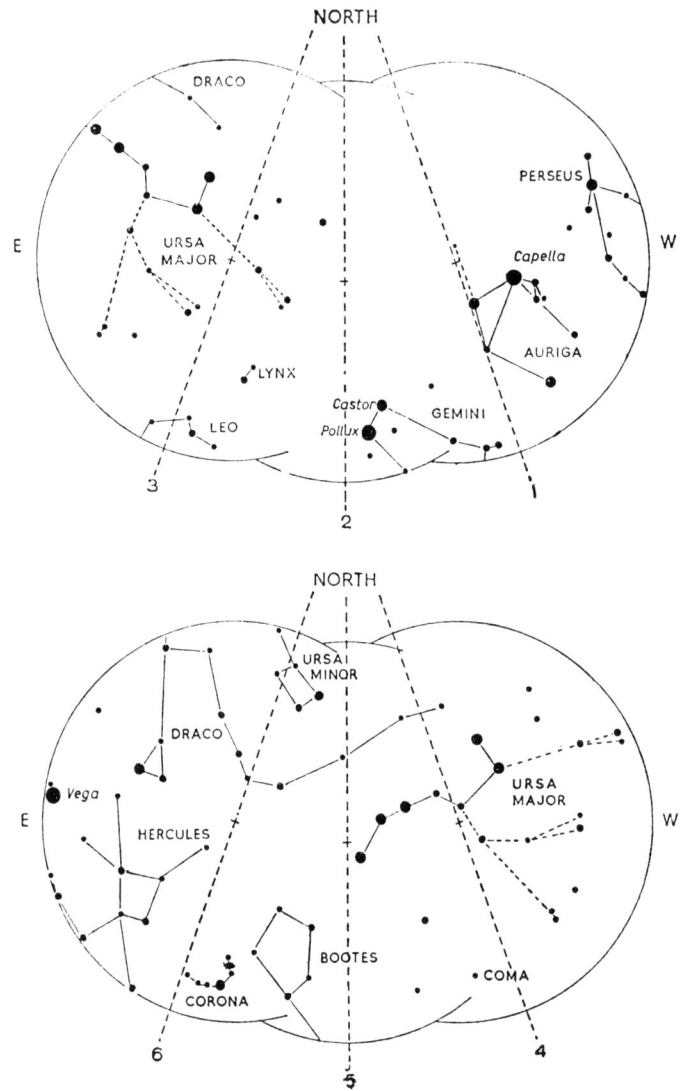

Northern Hemisphere Overhead Stars

NORTHERN STAR CHARTS

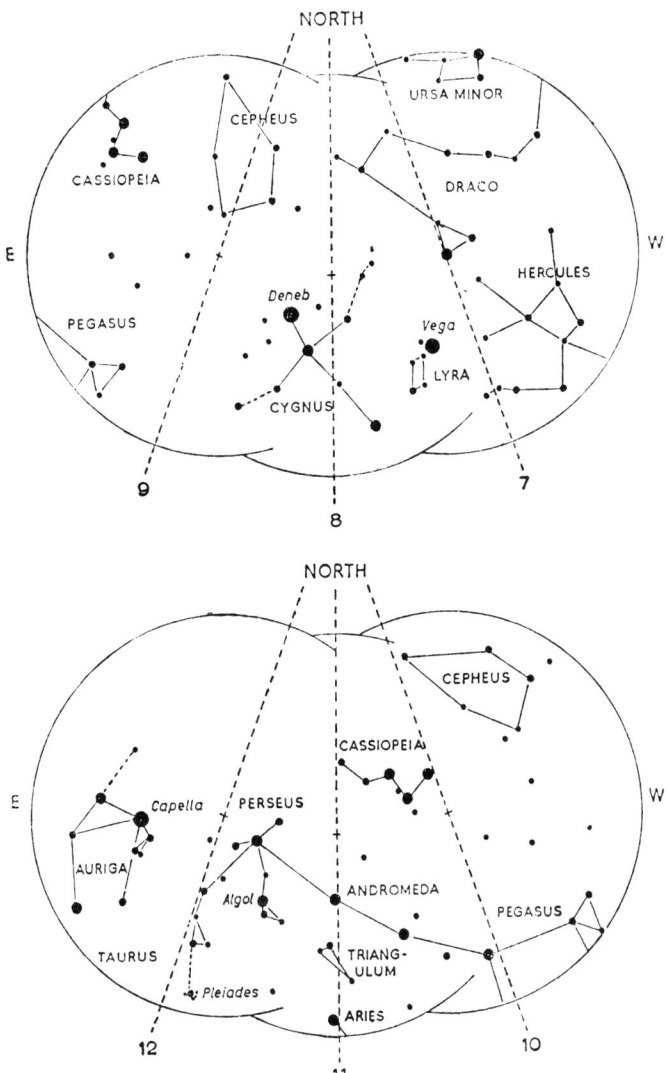

Northern Hemisphere Overhead Stars

1L

October 6 at 5^h	October 21 at 4^h
November 6 at 3^h	November 21 at 2^h
December 6 at 1^h	December 21 at midnight
January 6 at 23^h	January 21 at 22^h
February 6 at 21^h	February 21 at 20^h

SOUTHERN STAR CHARTS

October 6 at 5ʰ October 21 at 4ʰ
November 6 at 3ʰ November 21 at 2ʰ
December 6 at 1ʰ December 21 at midnight
January 6 at 23ʰ January 21 at 22ʰ
February 6 at 21ʰ February 21 at 20ʰ

1R

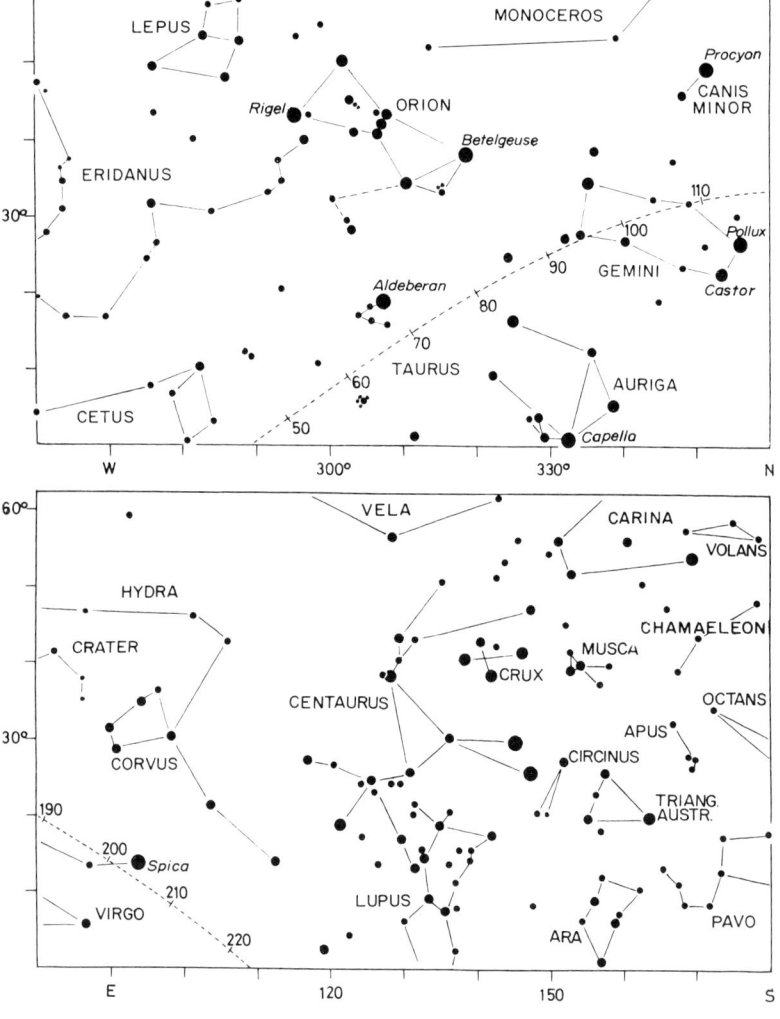

SOUTHERN STAR CHARTS

November 6 at 5ʰ   November 21 at 4ʰ
December 6 at 3ʰ   December 21 at 2ʰ
January 6 at 1ʰ   January 21 at midnight
February 6 at 23ʰ February 21 at 22ʰ
March 6 at 21ʰ   March 21 at 20ʰ

2R

3L

January 6 at 3ʰ	January 21 at 2ʰ
February 6 at 1ʰ	February 21 at midnight
March 6 at 23ʰ	March 21 at 22ʰ
April 6 at 21ʰ	April 21 at 20ʰ
May 6 at 19ʰ	May 21 at 18ʰ

SOUTHERN STAR CHARTS

January 6 at 3ʰ   January 21 at 2ʰ
February 6 at 1ʰ   February 21 at midnight
March 6 at 23ʰ   March 21 at 22ʰ
April 6 at 21ʰ   April 21 at 20ʰ
May 6 at 19ʰ   May 21 at 18ʰ

3R

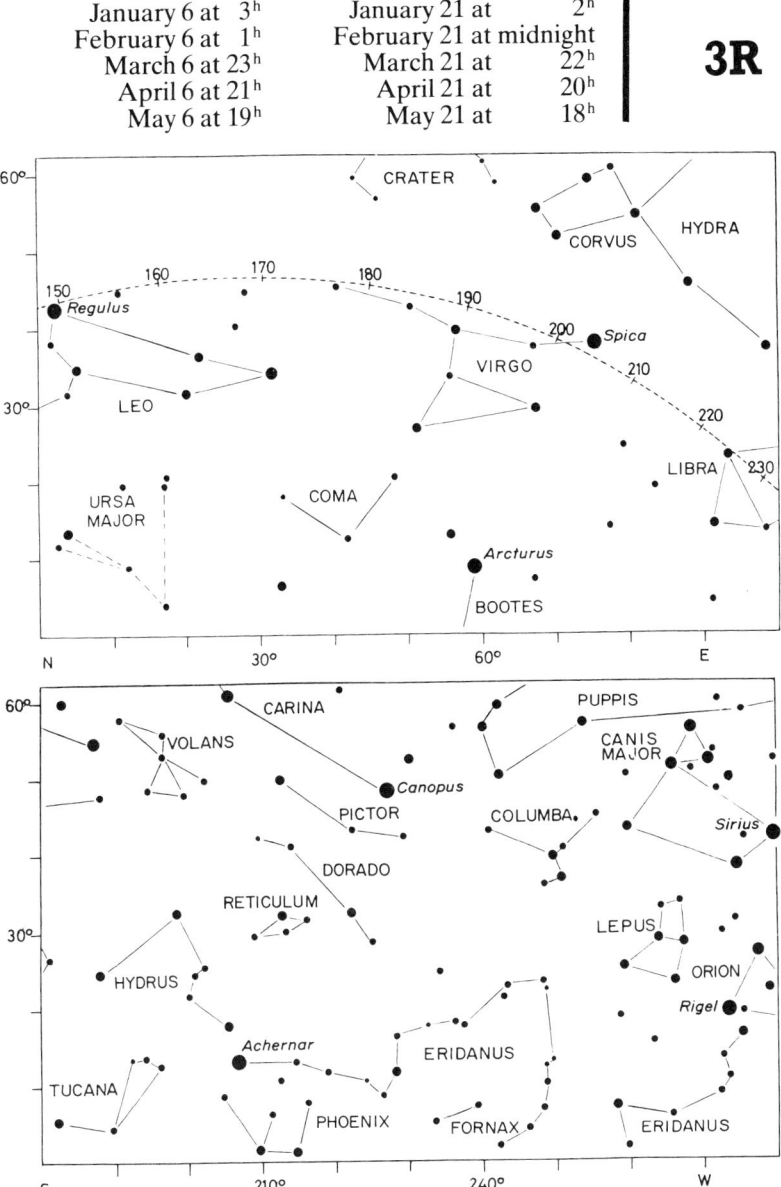

4L

February 6 at	3ʰ	February 21 at	2ʰ
March 6 at	1ʰ	March 21 at	midnight
April 6 at	23ʰ	April 21 at	22ʰ
May 6 at	21ʰ	May 21 at	20ʰ
June 6 at	19ʰ	June 21 at	18ʰ

SOUTHERN STAR CHARTS

February 6 at 3ʰ	February 21 at 2ʰ
March 6 at 1ʰ	March 21 at midnight
April 6 at 23ʰ	April 21 at 22ʰ
May 6 at 21ʰ	May 21 at 20ʰ
June 6 at 19ʰ	June 21 at 18ʰ

4R

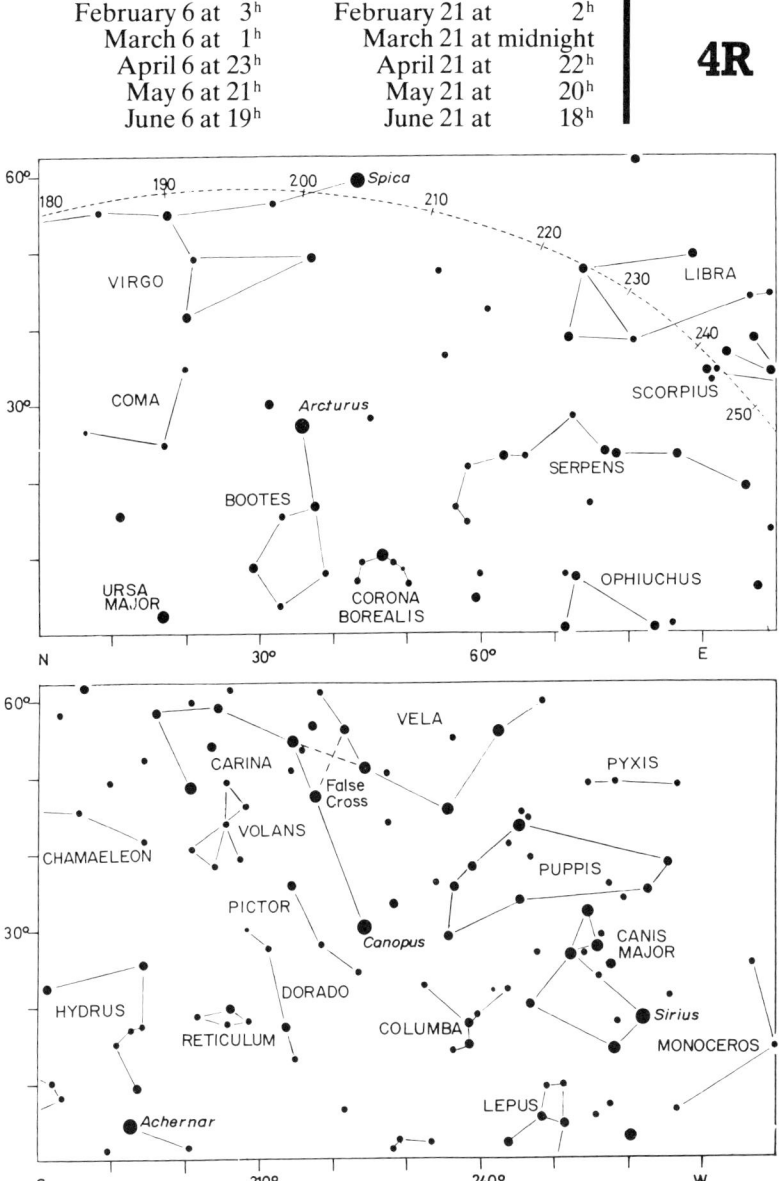

5L

March 6 at 3ʰ	March 21 at 2ʰ
April 6 at 1ʰ	April 21 at midnight
May 6 at 23ʰ	May 21 at 22ʰ
June 6 at 21ʰ	June 21 at 20ʰ
July 6 at 19ʰ	July 21 at 18ʰ

SOUTHERN STAR CHARTS

March 6 at 3ʰ   March 21 at 2ʰ
April 6 at 1ʰ   April 21 at midnight
May 6 at 23ʰ   May 21 at 22ʰ
June 6 at 21ʰ   June 21 at 20ʰ
July 6 at 19ʰ   July 21 at 18ʰ

5R

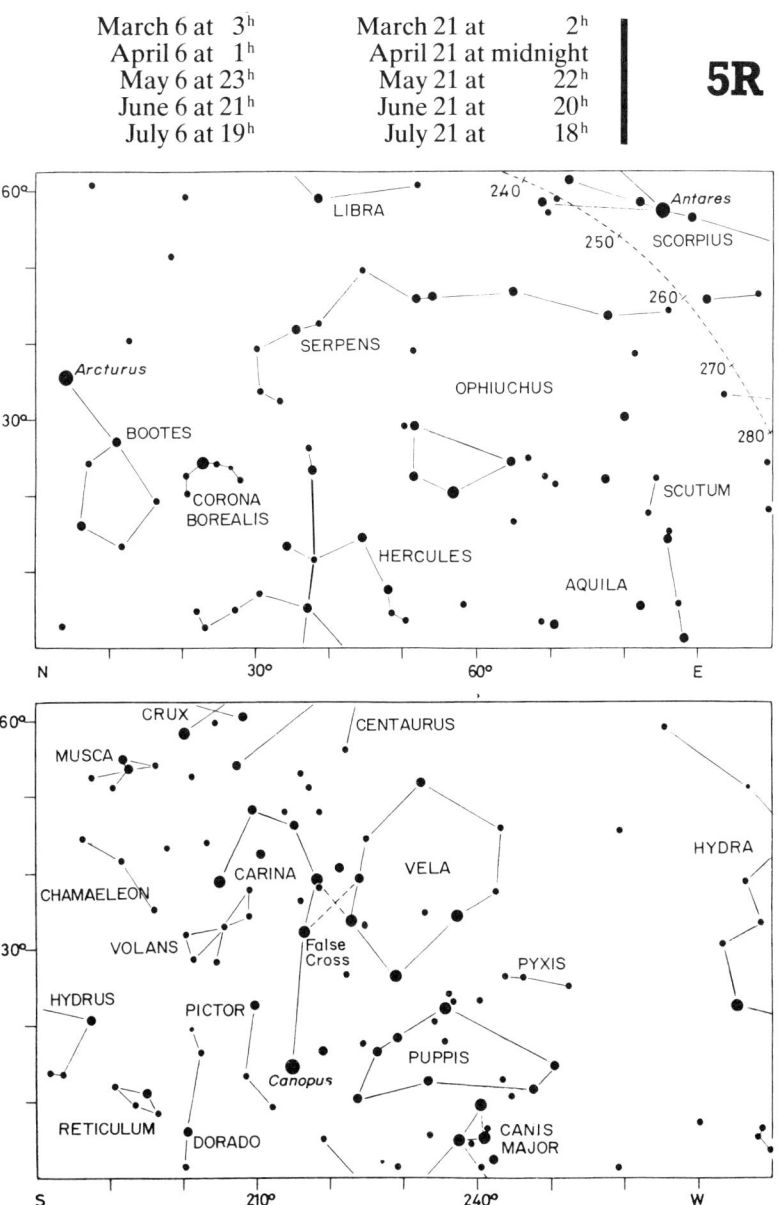

1988 YEARBOOK OF ASTRONOMY

6L

March 6 at 5^h	March 21 at 4^h
April 6 at 3^h	April 21 at 2^h
May 6 at 1^h	May 21 at midnight
June 6 at 23^h	June 21 at 22^h
July 6 at 21^h	July 21 at 20^h

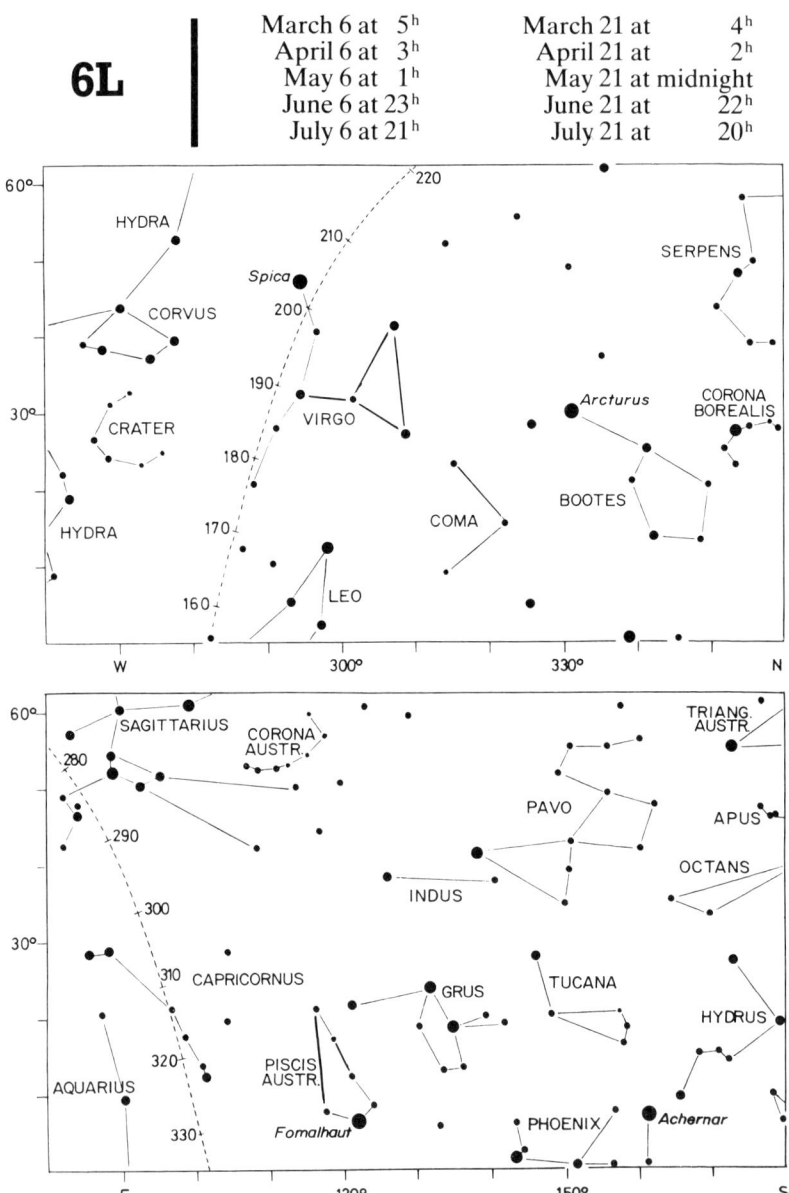

SOUTHERN STAR CHARTS

March 6 at 5ʰ   March 21 at 4ʰ
April 6 at 3ʰ   April 21 at 2ʰ
May 6 at 1ʰ   May 21 at midnight
June 6 at 23ʰ   June 21 at 22ʰ
July 6 at 21ʰ   July 21 at 20ʰ

6R

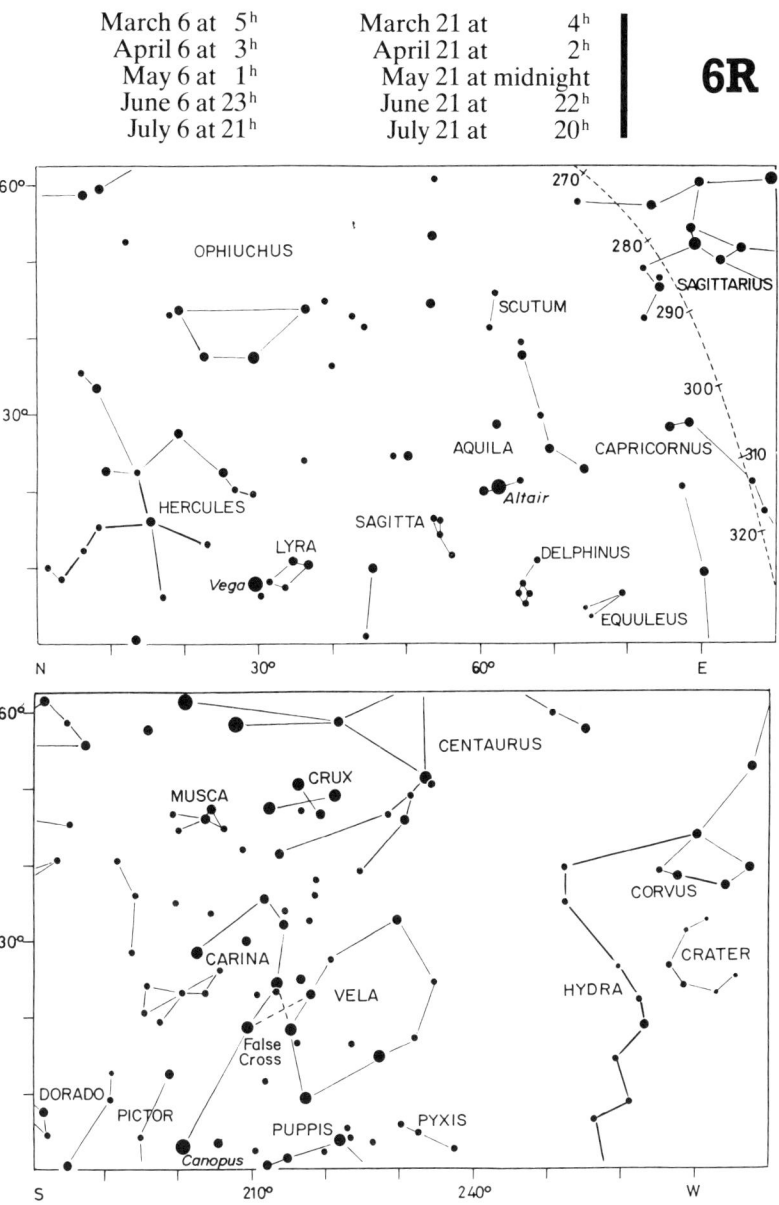

51

7L

April 6 at	5^h	April 21 at	4^h
May 6 at	3^h	May 21 at	2^h
June 6 at	1^h	June 21 at midnight	
July 6 at	23^h	July 21 at	22^h
August 6 at	21^h	August 21 at	20^h

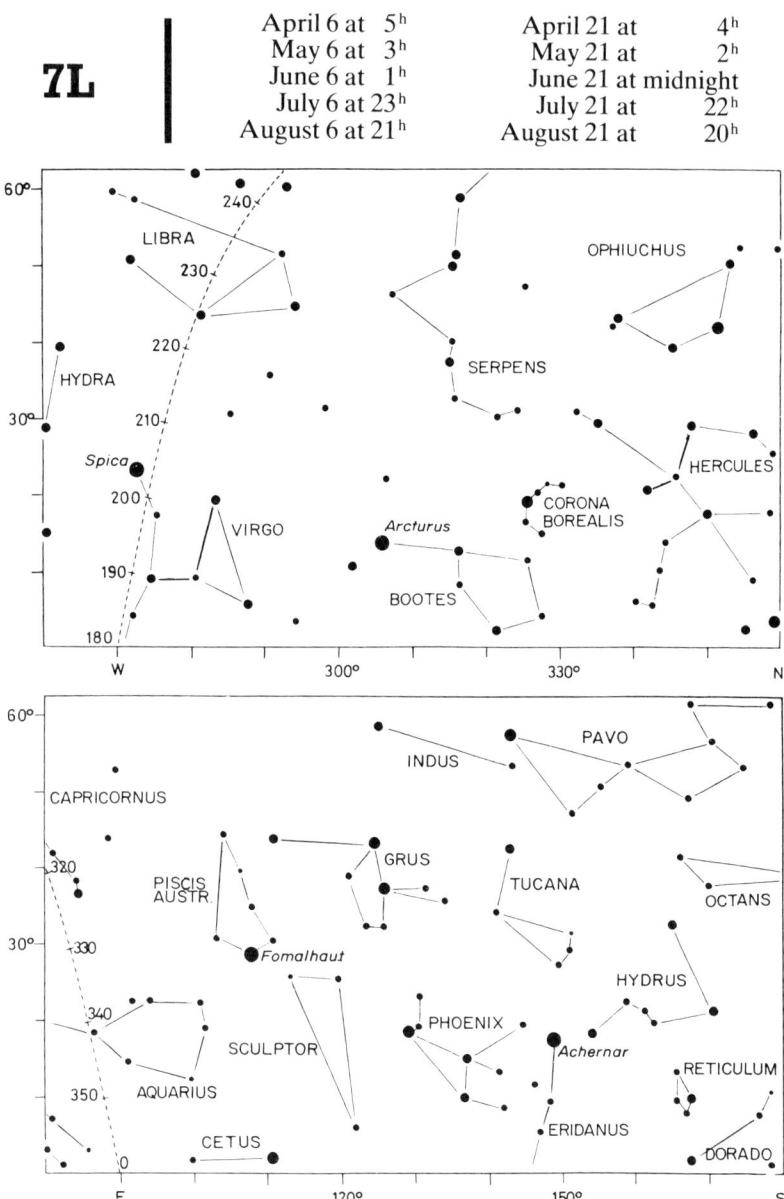

SOUTHERN STAR CHARTS

April 6 at 5ʰ
May 6 at 3ʰ
June 6 at 1ʰ
July 6 at 23ʰ
August 6 at 21ʰ

April 21 at 4ʰ
May 21 at 2ʰ
June 21 at midnight
July 21 at 22ʰ
August 21 at 20ʰ

7R

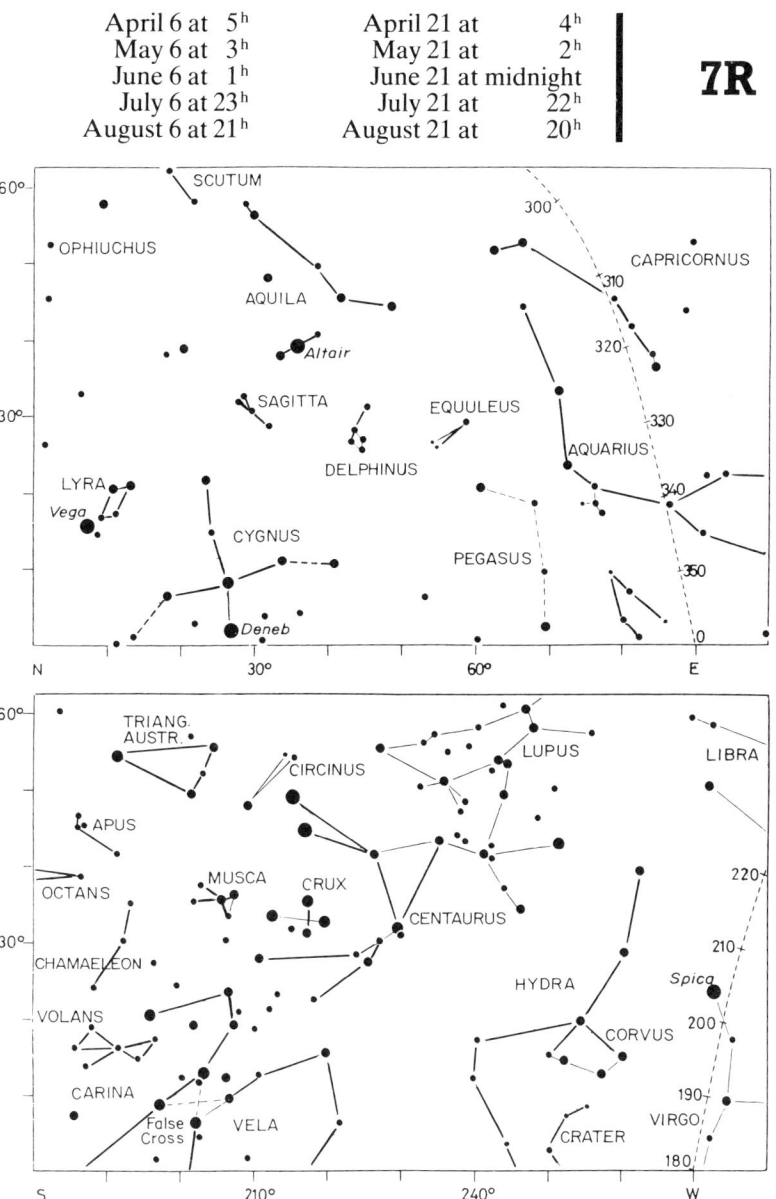

8L

May 6 at 5ʰ	May 21 at 4ʰ
June 6 at 3ʰ	June 21 at 2ʰ
July 6 at 1ʰ	July 21 at midnight
August 6 at 23ʰ	August 21 at 22ʰ
September 6 at 21ʰ	September 21 at 20ʰ

SOUTHERN STAR CHARTS

May 6 at 5ʰ   May 21 at 4ʰ
June 6 at 3ʰ   June 21 at 2ʰ
July 6 at 1ʰ   July 21 at midnight
August 6 at 23ʰ August 21 at 22ʰ
September 6 at 21ʰ September 21 at 20ʰ

8R

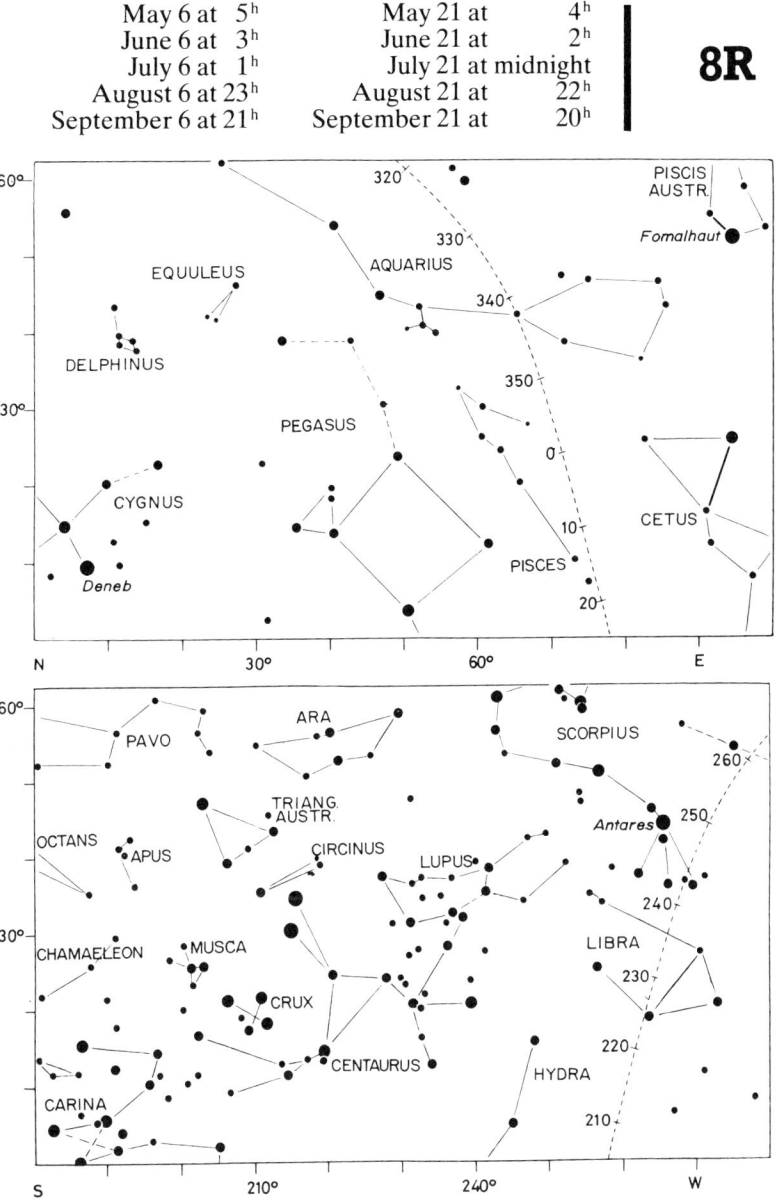

1988 YEARBOOK OF ASTRONOMY

9L

June 6 at 5^h	June 21 at 4^h
July 6 at 3^h	July 21 at 2^h
August 6 at 1^h	August 21 at midnight
September 6 at 23^h	September 21 at 22^h
October 6 at 21^h	October 21 at 20^h

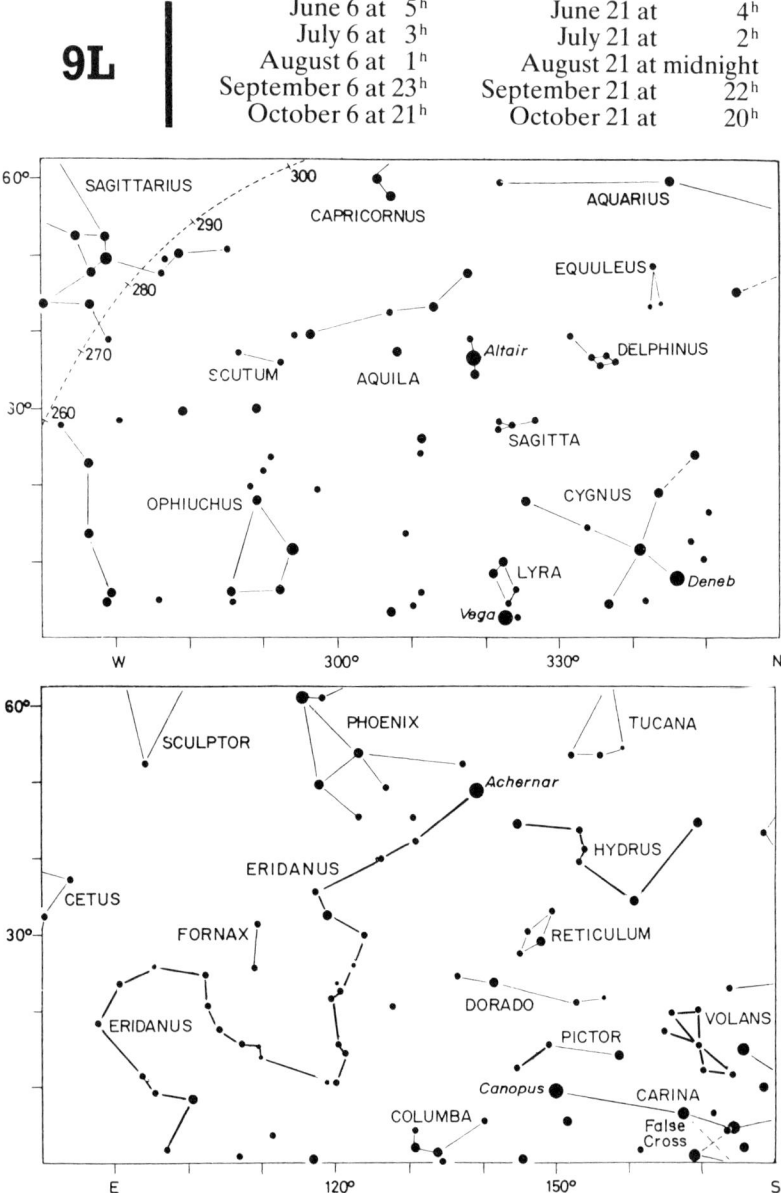

SOUTHERN STAR CHARTS

June 6 at 5ʰ   June 21 at 4ʰ
July 6 at 3ʰ   July 21 at 2ʰ
August 6 at 1ʰ   August 21 at midnight
September 6 at 23ʰ   September 21 at 22ʰ
October 6 at 21ʰ   October 21 at 20ʰ

9R

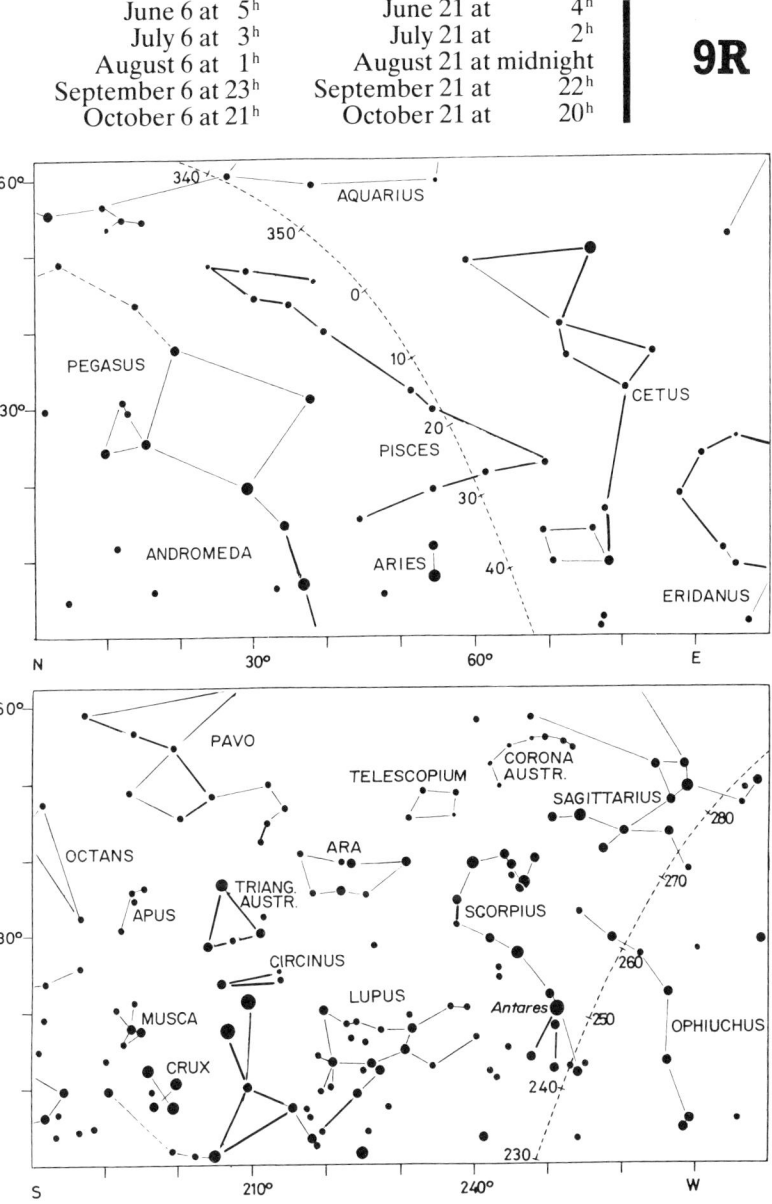

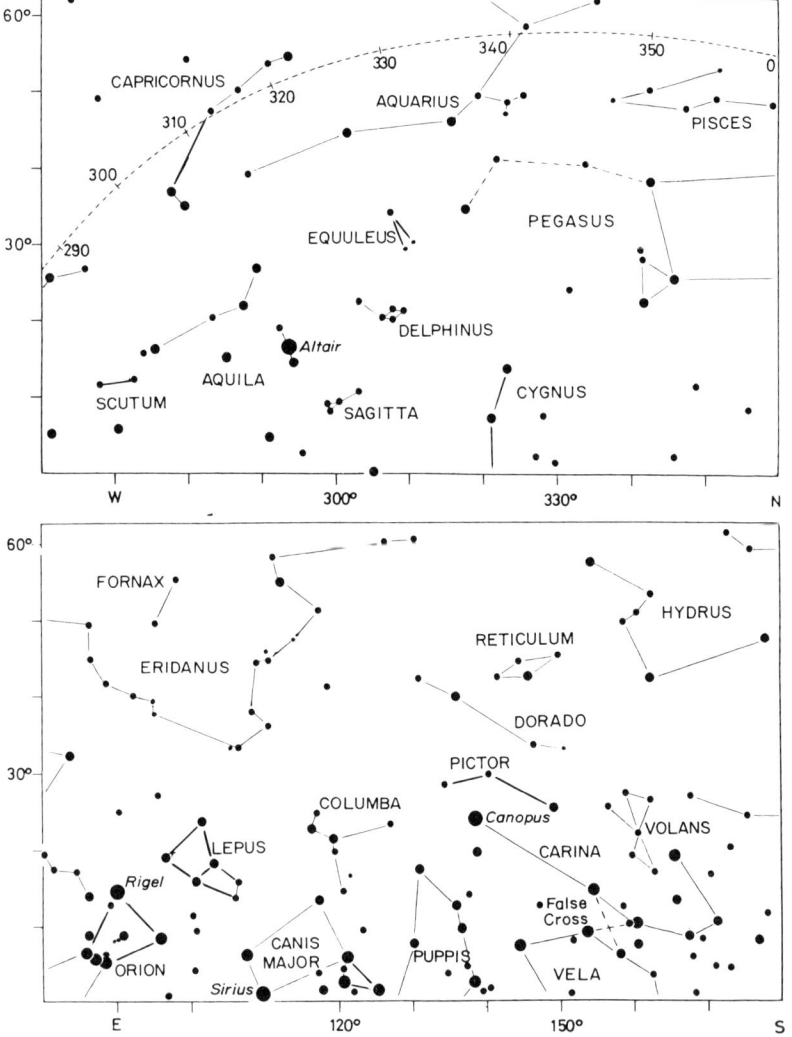

SOUTHERN STAR CHARTS

July 6 at 5^h	July 21 at 4^h
August 6 at 3^h	August 21 at 2^h
September 6 at 1^h	September 21 at midnight
October 6 at 23^h	October 21 at 22^h
November 6 at 21^h	November 21 at 20^h

10R

11L

August 6 at 5ʰ	August 21 at 4ʰ
September 6 at 3ʰ	September 21 at 2ʰ
October 6 at 1ʰ	October 21 at midnight
November 6 at 23ʰ	November 21 at 22ʰ
December 6 at 21ʰ	December 21 at 20ʰ

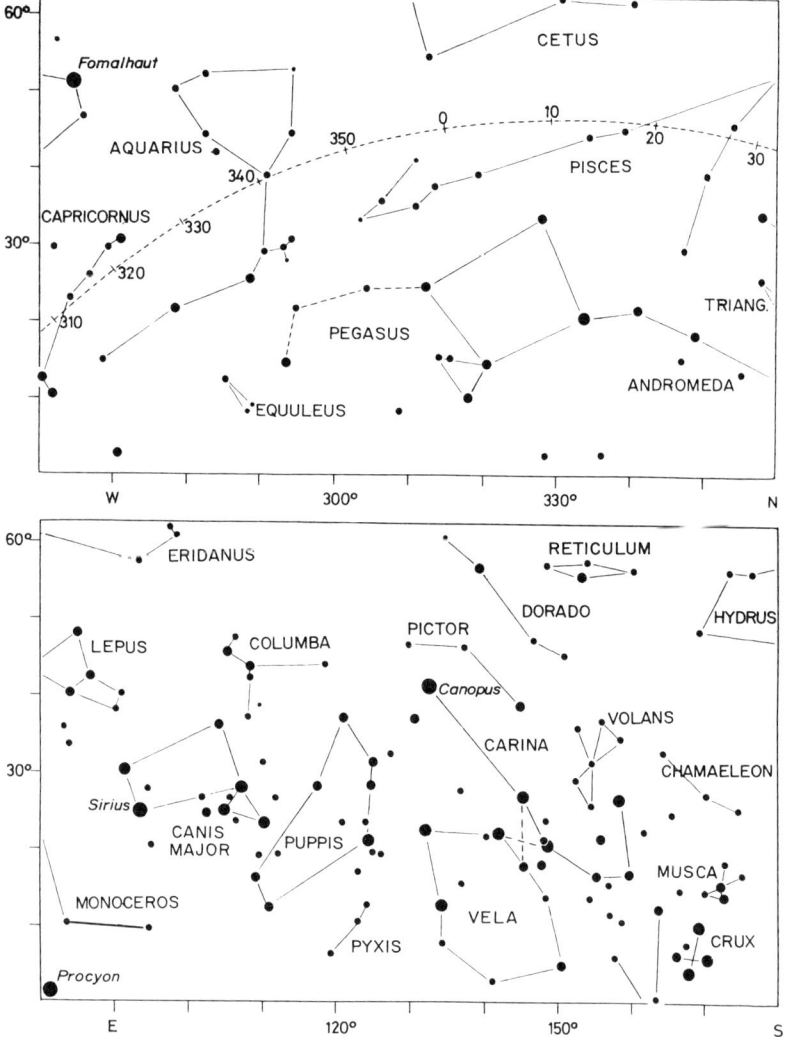

SOUTHERN STAR CHARTS

August 6 at 5ʰ
September 6 at 3ʰ
October 6 at 1ʰ
November 6 at 23ʰ
December 6 at 21ʰ

August 21 at 4ʰ
September 21 at 2ʰ
October 21 at midnight
November 21 at 22ʰ
December 21 at 20ʰ

11R

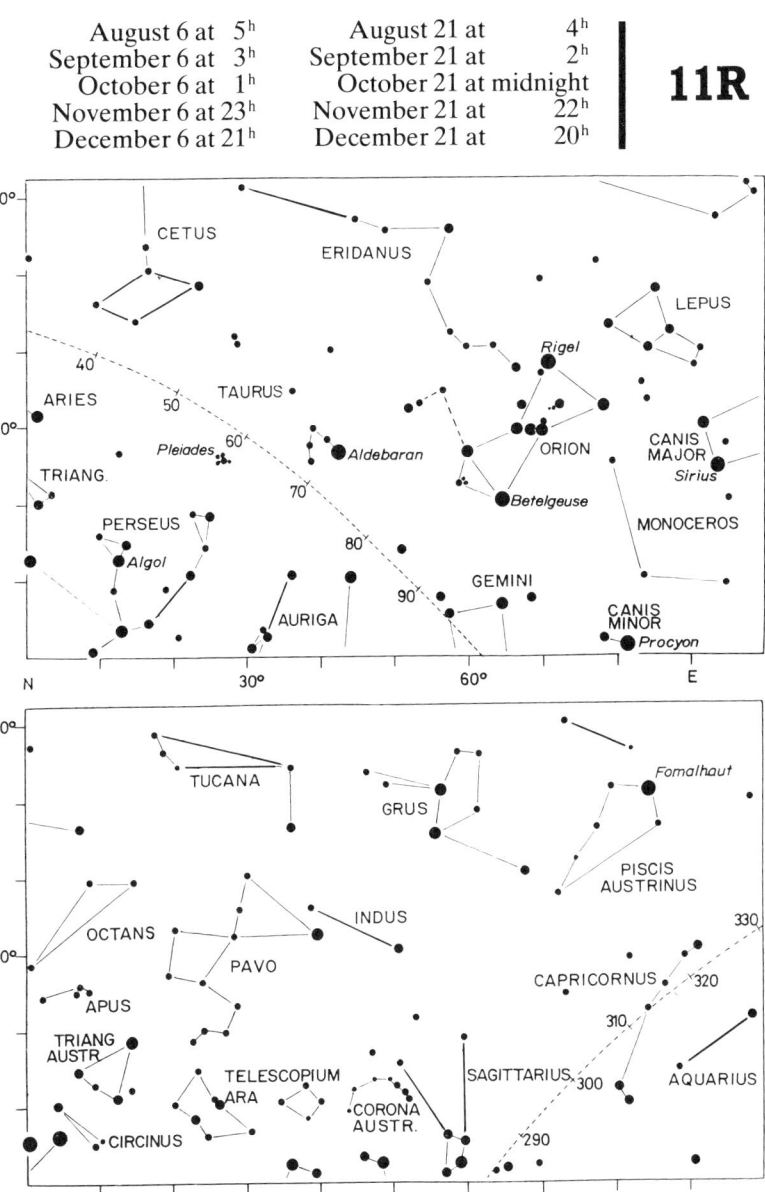

12L

September 6 at	5^h	September 21 at	4^h
October 6 at	3^h	October 21 at	2^h
November 6 at	1^h	November 21 at midnight	
December 6 at	23^h	December 21 at	22^h
January 6 at	21^h	January 21 at	20^h

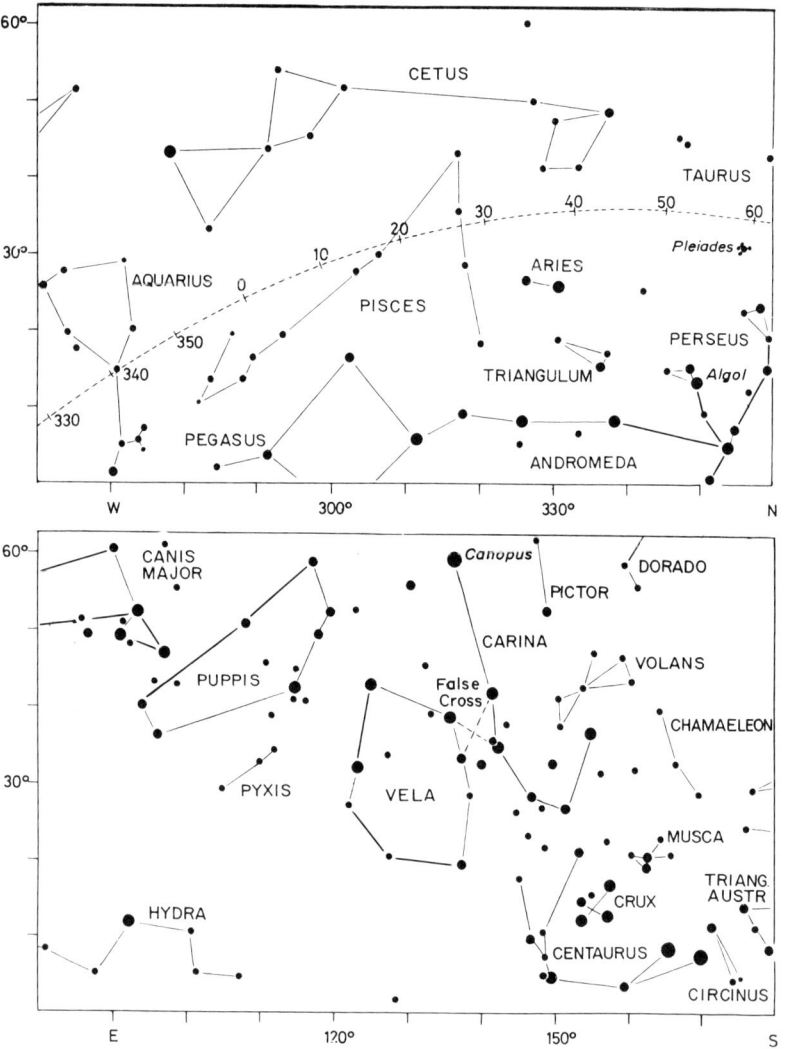

SOUTHERN STAR CHARTS

September 6 at 5ʰ September 21 at 4ʰ
October 6 at 3ʰ October 21 at 2ʰ
November 6 at 1ʰ November 21 at midnight
December 6 at 23ʰ December 21 at 22ʰ
January 6 at 21ʰ January 21 at 20ʰ

12R

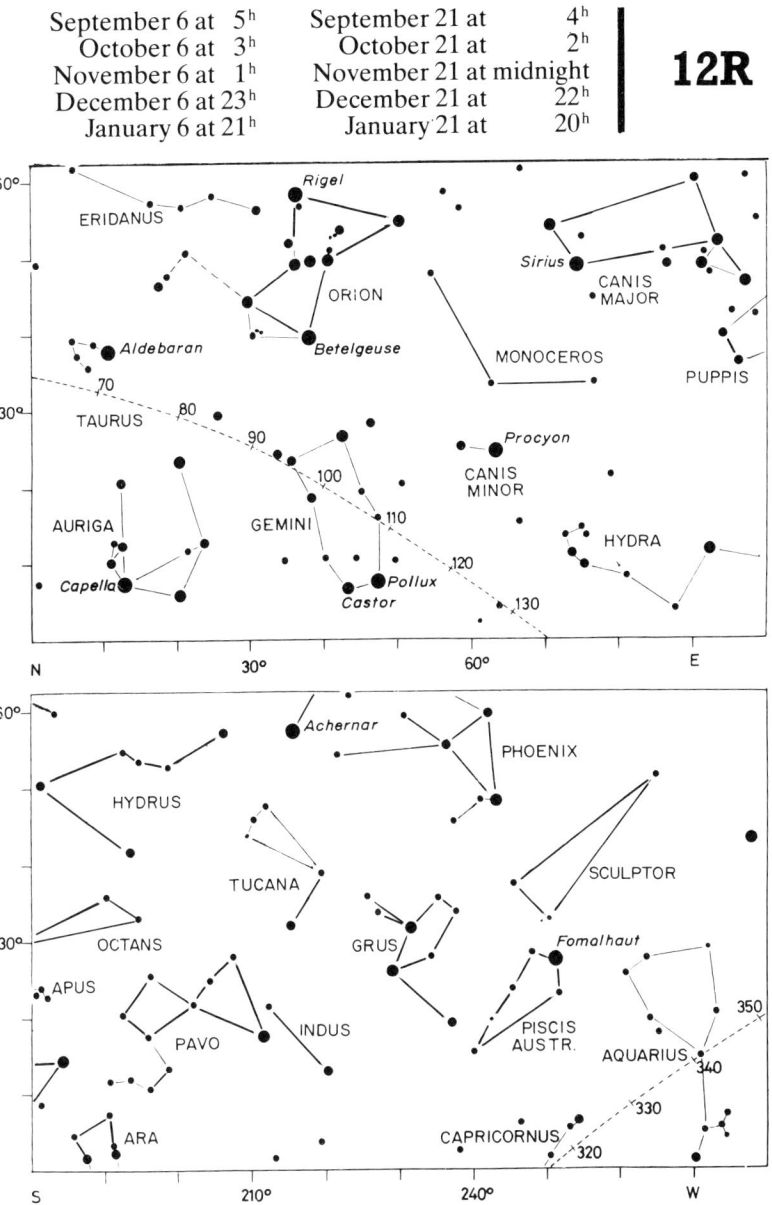

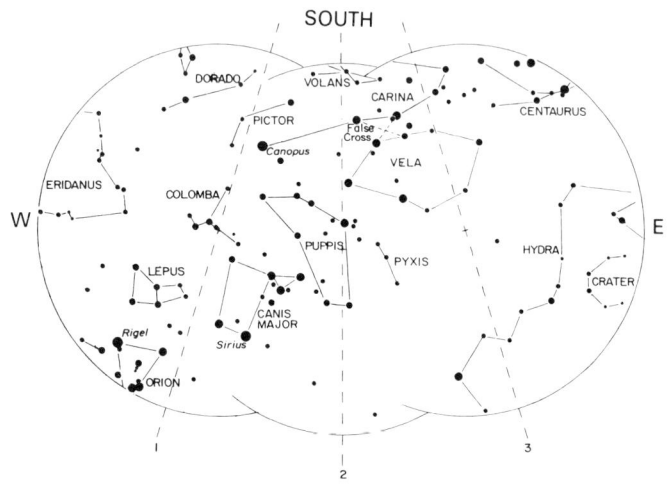
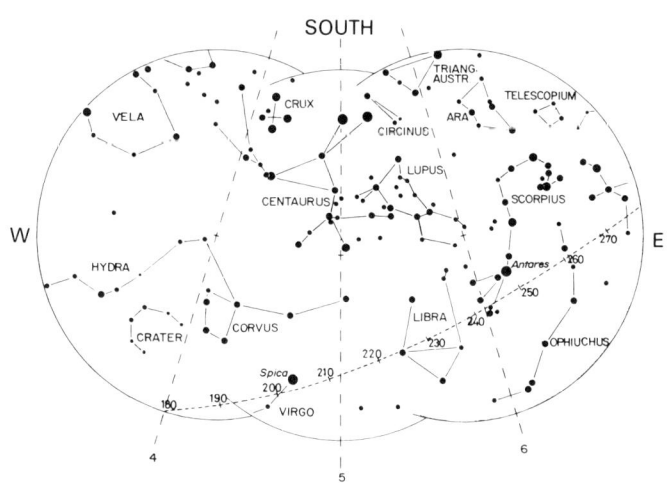

Southern Hemisphere Overhead Stars

SOUTHERN STAR CHARTS

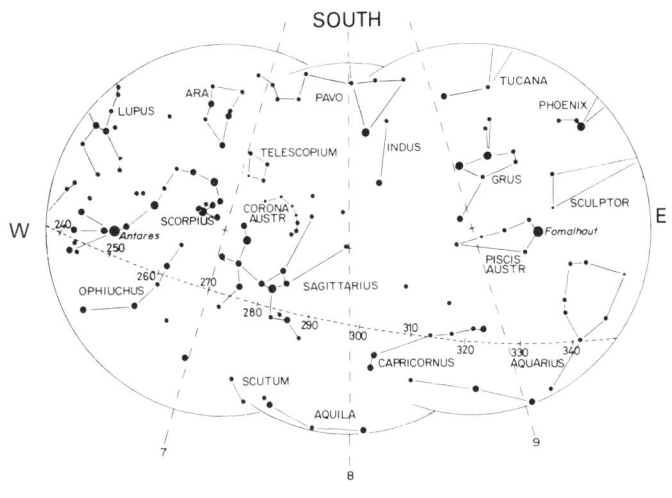

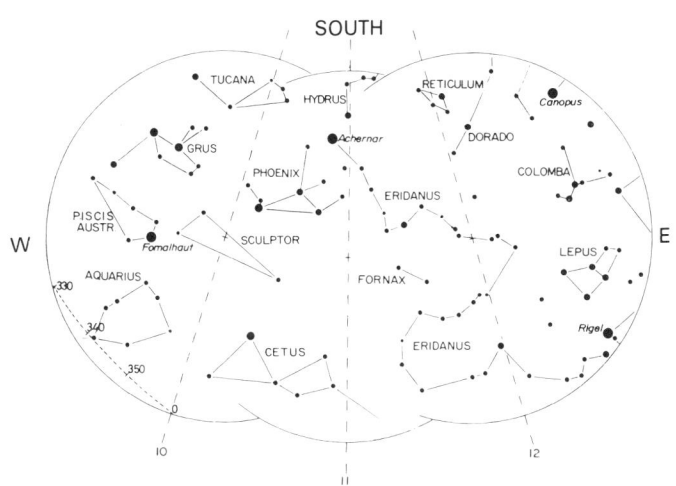

Southern Hemisphere Overhead Stars

The Planets and the Ecliptic

The paths of the planets about the Sun all lie close to the plane of the ecliptic, which is marked for us in the sky by the apparent path of the Sun among the stars, and is shown on the star charts by a broken line. The Moon and planets will always be found close to this line, never departing from it by more than about 7 degrees. Thus the planets are most favourably placed for observation when the ecliptic is well displayed, and this means that it is should be as high in the sky as possible. This avoids the difficulty of finding a clear horizon, and also overcomes the problem of atmospheric absorption, which greatly reduces the light of the stars. Thus a star at an altitude of 10 degrees suffers a loss of 60 per cent of its light, which corresponds to a whole magnitude; at an altitude of only 4 degrees, the loss may amount to two magnitudes.

The position of the ecliptic in the sky is therefore of great importance, and since it is tilted at about $23\frac{1}{2}$ degrees to the Equator, it is only at certain times of the day or year that it is displayed to the best advantage. It will be realized that the Sun (and therefore the ecliptic) is at its highest in the sky at noon in midsummer, and at its lowest at noon in midwinter. Allowing for the daily motion of the sky, these times lead to the fact that the ecliptic is highest at midnight in winter, at sunset in the spring, at noon in summer and at sunrise in the autumn. Hence these are the best times to see the planets. Thus, if Venus is an evening object, in the western sky after sunset, it will be seen to best advantage if this occurs in the spring, when the ecliptic is high in the sky and slopes down steeply to the horizon. This means that the planet is not only higher in the sky, but will remain for a much longer period above the horizon. For similar reasons, a morning object will be seen at its best on autumn mornings before sunrise, when the ecliptic is high in the east. The outer planets, which can come to opposition (i.e. opposite the Sun), are best seen when opposition occurs in the winter months, when the ecliptic is high in the sky at midnight.

The seasons are reversed in the Southern Hemisphere, spring beginning at the September Equinox, when the Sun crosses the Equator on its way south, summer beginning at the December Solstice, when the Sun is highest in the southern sky, and so on. Thus, the times when the ecliptic is highest in the sky, and therefore best placed for observing the planets, may be summarized as follows:

	Midnight	*Sunrise*	*Noon*	*Sunset*
Northern lats.	December	September	June	March
Southern lats.	June	March	December	September

In addition to the daily rotation of the celestial sphere from east to west, the planets have a motion of their own among the stars. The apparent movement is generally *direct,* i.e. to the east, in the direction of increasing longitude, but for a certain period (which depends on the distance of the planet) this apparent motion is reversed. With the outer planets this *retrograde* motion occurs about the time of opposition. Owing to the different inclination of the orbits of these planets, the actual effect is to cause the apparent path to form a loop, or sometimes an S-shaped curve. The same effect is present in the motion of the inferior planets, Mercury and Venus, but it is not so obvious, since it always occurs at the time of inferior conjunction.

The inferior planets, Mercury and Venus, move in smaller orbits than that of the Earth, and so are always seen near the Sun. They are most obvious at the times of greatest angular distance from the Sun (greatest elongation), which may reach 28 degrees for Mercury, or 47 degrees for Venus. They are seen as evening objects in the western sky after sunset (at eastern elongations) or as morning objects in the eastern sky before sunrise (at western elongations). The succession of phenomena, conjunctions and elongations, always follows the same order, but the intervals between them are not equal. Thus, if either planet is moving round the far side of its orbit its motion will be to the east, in the same direction in which the Sun appears to be moving. It therefore takes much longer for the planet to overtake the Sun – that is, to come to superior conjunction – than it does when moving round to inferior conjunction, between Sun and Earth. The intervals given in the following table are average values; they remain fairly constant in the case of Venus, which travels in an almost circular orbit. In the case of Mercury, however, conditions vary widely because of the great eccentricity and inclination of the planet's orbit.

		Mercury	Venus
Inferior conj.	to Elongation West	22 days	72 days
Elongation West	to Superior conj.	36 days	220 days
Superior conj.	to Elongation East	36 days	220 days
Elongation East	to Inferior conj.	22 days	72 days

The greatest brilliancy of Venus always occurs about 36 days before or after inferior conjunction. This will be about a month *after* greatest eastern elongation (as an evening object), or a month *before* greatest western elongation (as a morning object). No such rule can be given for Mercury, because its distance from the Earth and the Sun can vary over a wide range.

Mercury is not likely to be seen unless a clear horizon is available. It is seldom seen as much as 10 degrees above the horizon in the twilight sky in northern latitudes, but this figure is often exceeded in the Southern Hemisphere. This favourable condition arises because the maximum elongation of 28 degrees can occur only when the planet is at aphelion (farthest from the Sun), and this point lies well south of the Equator. Northern observers must be content with smaller elongations, which may be as little as 18 degrees at perihelion. In general, it may be said that the most favourable times for seeing Mercury as an evening object will be in spring, some days before greatest eastern elongation; in autumn, it may be seen as a morning object some days after greatest western elongation.

Venus is the brightest of the planets and may be seen on occasions in broad daylight. Like Mercury, it is alternately a morning and an evening object, and it will be highest in the sky when it is a morning object in autumn, or an evening object in spring. The phenomena of Venus given in the table above can occur only in the months of January, April, June, August and November, and it will be realized that they do not all lead to favourable apparitions of the planet. In fact, Venus is to be seen at its best as an evening object in northern latitudes when eastern elongations occurs in June. The planet is then well north of the Sun in the preceding spring months, and is a brilliant object in the evening sky over a long period. In the Southern Hemisphere a November elongation is best. For similar reasons, Venus gives a prolonged display as a morning object in the months following western elongation in November (in northern latitudes) or in June (in the Southern Hemisphere).

The superior planets, which travel in orbits larger than that of the Earth, differ from Mercury and Venus in that they can be seen opposite the Sun in the sky. The superior planets are morning objects after conjunction with the Sun, rising earlier each day until they come to opposition. They will then be nearest to the Earth (and therefore at their brightest), and will then be on the meridian at midnight, due south in northern latitudes, but due north in the Southern Hemisphere. After opposition they are evening objects, setting earlier each evening until they set in the west with the Sun at the next conjunction. The change in brightness about the time of opposition is most noticeable in the case of Mars, whose distance from Earth can vary considerably and rapidly. The other superior planets are at such great distances that there is very little change in brightness from one opposition to another. The effect of altitude is, however, of some importance, for at a December opposition in northern latitudes the planets will be among the stars of Taurus or Gemini, and can then be at an altitude of more than 60 degrees in southern England. At a summer opposition, when the planet is in Sagittarius, it may only rise to about 15 degrees above the southern horizon, and so makes a less impressive appearance. In the Southern Hemisphere, the reverse conditions apply; a June opposition being the best, with the planet in Sagittarius at an altitude which can reach 80 degrees above the northern horizon for observers in South Africa.

Mars, whose orbit is appreciably eccentric, comes nearest to the Earth at an opposition at the end of August. It may then be brighter even than Jupiter, but rather low in the sky in Aquarius for northern observers, though very well placed for those in southern latitudes. These favourable oppositions occur every fifteen or seventeen years (1956, 1971, 1988, 2003) but in the Northern Hemisphere the planet is probably better seen at an opposition in the autumn or winter months, when it is higher in the sky. Oppositions of Mars occur at an average interval of 780 days, and during this time the planet makes a complete circuit of the sky.

Jupiter is always a bright planet, and comes to opposition a month later each year, having moved, roughly speaking, from one Zodiacal constellation to the next.

Saturn moves much more slowly than Jupiter, and may remain in the same constellation for several years. The brightness of Saturn depends on the aspects of its rings, as well as on the distance from Earth and Sun. The rings have been inclined towards the Earth and Sun at quite a small angle, and are opening again after being seen edge-on in 1980. The next passage of both the Earth and the Sun through the ring-plane will not occur until 1995.

Uranus, Neptune, and *Pluto* are hardly likely to attract the attention of observers without adequate instruments.

Phases of the Moon 1988

New Moon	First Quarter	Full Moon	Last Quarter
d h m	d h m	d h m	d h m
		Jan. 4 01 40	Jan. 12 07 04
Jan. 19 05 26	Jan. 25 21 53	Feb. 2 20 51	Feb. 10 23 01
Feb. 17 15 54	Feb. 24 12 15	Mar. 3 16 01	Mar. 11 10 56
Mar. 18 02 02	Mar. 25 04 41	Apr. 2 09 21	Apr. 9 19 21
Apr. 16 12 00	Apr. 23 22 32	May 1 23 41	May 9 01 23
May 15 22 11	May 23 16 49	May 31 10 53	June 7 06 21
June 14 09 14	June 22 10 23	June 29 19 46	July 6 11 36
July 13 21 53	July 22 02 14	July 29 03 25	Aug. 4 18 22
Aug. 12 12 31	Aug. 20 15 51	Aug. 27 10 56	Sept. 3 03 50
Sept. 11 04 49	Sept. 19 03 18	Sept. 25 19 07	Oct. 2 16 58
Oct. 10 21 49	Oct. 18 13 01	Oct. 25 04 35	Nov. 1 10 11
Nov. 9 14 20	Nov. 16 21 35	Nov. 23 15 53	Dec. 1 06 49
Dec. 9 05 36	Dec. 16 05 40	Dec. 23 05 29	Dec. 31 04 57

All times are G.M.T.

Reproduced, with permission, from data supplied by the Science and Engineering Research Council.

Longitudes of the Sun, Moon and Planets in 1988

DATE		Sun°	Moon°	Venus°	Mars°	Jupiter°	Saturn°
January	6	285	126	318	238	21	266
	21	301	325	336	248	22	268
February	6	317	170	356	259	24	269
	21	332	18	13	269	27	271
March	6	346	191	30	278	29	271
	21	1	40	46	289	33	272
April	6	17	239	62	299	36	273
	21	32	88	76	309	40	272
May	6	46	276	86	319	43	272
	21	61	121	90	329	47	271
June	6	76	329	86	339	51	270
	21	90	164	78	348	54	269
July	6	105	8	74	356	57	268
	21	119	196	79	3	60	267
August	6	134	59	89	9	62	266
	21	149	242	102	11	64	266
September	6	164	107	118	11	66	266
	21	179	292	135	7	66	266
October	6	193	140	152	3	66	267
	21	208	330	169	0	65	268
November	6	224	183	188	0	63	269
	21	239	23	207	3	61	271
December	6	255	216	225	8	59	273
	21	270	61	244	15	58	274

Longitude of *Uranus* 269°
Neptune 279°

Moon: Longitude of ascending node
Jan. 1: 357°   Dec. 31: 338°

LONGTITUDES OF THE SUN, MOON AND PLANETS IN 1988

Mercury moves so quickly among the stars that it is not possible to indicate its position on the star charts at a convenient interval. The monthly notes must be consulted for the best times at which the planet may be seen.

The positions of the other planets are given in the table on the previous page. This gives the apparent longitudes on dates which correspond to those of the star charts, and the position of the planet may at once be found near the ecliptic at the given longitude.

Examples
In the southern hemisphere two planets are seen in the eastern morning sky in February. Identify them.

> The southern star charts 4L and 4R shows the eastern sky at February 6^d3^h and shows longitudes $180° - 280°$. Reference to the table opposite gives the longitude of Mars as 269° and that of Saturn as 271°, on February 21. Thus these planets are found to the north-north east of Antares and the one with the slightly reddish tint is Mars.

The positions of the Sun and Moon can be plotted on the star maps in the same manner as for the planets. The average daily motion of the Sun is 1°, and of the Moon 13°. For the Moon an indication of its position relative to the ecliptic may be obtained from a consideration of its longitude relative to that of the ascending node. The latter changes only slowly during the year as will be seen from the values given on the opposite page. Let us call the difference in longitude of Moon-node, d. Then if d=0°, 180° or 360° the Moon is on the ecliptic. If d=90° the Moon is 5° north of the ecliptic and if d=270° the Moon is 5° south of the ecliptic.

On May 6 the Moon's longitude is given as 276° and the longitude of the node is found by interpolation to be about 345°. Thus d=257° and the Moon is about 5° south of the ecliptic. Its position may be plotted on northern star charts 6L and 7L: and southern star charts 4L, 5L, 6R, 9L, 9R and 10R.

Some Events in 1988

ECLIPSES
In 1988 there will be four eclipses, two of the Sun and two of the Moon.

March 3: partial eclipse of the Moon – Alaska, Arctic regions, Pacific Ocean, Australiasia, Asia, E. Africa, N.E. Europe.

March 17-18: total eclipse of the Sun – E. Asia, Indonesia, N.W. Australia, New Guinea, Micronesia, extreme N.W. of N. America, W. Hawaiian Islands.

August 27: partial eclipse of the Moon – N. America (except E.), Central America, W. of S. America, Antarctica, Pacific Ocean, Australasia, E. Asia.

September 11: annular eclipse of the Sun – extreme E. Africa, S. Asia, Indonesia, Australia (except extreme N.E.), New Zealand, part of Antarctica.

THE PLANETS
Mercury may be seen more easily from northern latitudes in the evenings about the time of greatest eastern elongation (January 26) and in the mornings around greatest western elongation (October 26). In the Southern Hemisphere the corresponding dates are March 8 (morning) and September 15 (evening).

Venus is visible in the evenings until May and in the mornings from July onwards.

Mars is a morning object until opposition in September when it is an evening object for the rest of the year.

Jupiter is at opposition on November 21.

Saturn is at opposition on June 20.

Uranus is at opposition on June 20.

Neptune is at opposition on June 30.

Pluto is at opposition on May 2.

MONTHLY NOTES 1988

January

Full Moon: January 4   *New Moon:* January 19

Earth is at perihelion (nearest to the Sun) on January 4 at a distance of 147 million kilometres.

Mercury attains its greatest eastern elongation (19°) from the Sun on January 26. It is visible as an evening object after the middle of the month (though observers in the latitudes of the British Isles are unlikely to sight it before January 18). This elusive planet may be glimpsed low above the south-western horizon at the end of evening civil twilight.

Venus is a brilliant evening object, magnitude −4.0. Observers in the British Isles will note that the period available for observation increases from two to three hours during the month as the planet moves northward in declination and increases its angular distance from the Sun.

Mars will be visible at some time during the hours of darkness during the whole of 1988. This month, Mars, magnitude +1.5, is visible in the south-eastern sky before the increasing morning twilight inhibits observation. During January Mars moves from Libra, through the northern part of Scorpius into Sagittarius, passing 5° N. of Antares on January 21. Figure 1 (overleaf) shows the path of Mars amongst the stars during the early part of the year.

Jupiter, magnitude −2.4, is an evening object in Pisces and its path amongst the stars is shown in Figure 5 (given with the monthly notes for July).

Saturn is rather too close to the Sun for observation early in the month. However, it should be visible low in the south-eastern sky before the onset of morning twilight inhibits observation, after the first week of the month for observers in the Southern Hemisphere, and about a week later for observers in northern temperate latitudes.

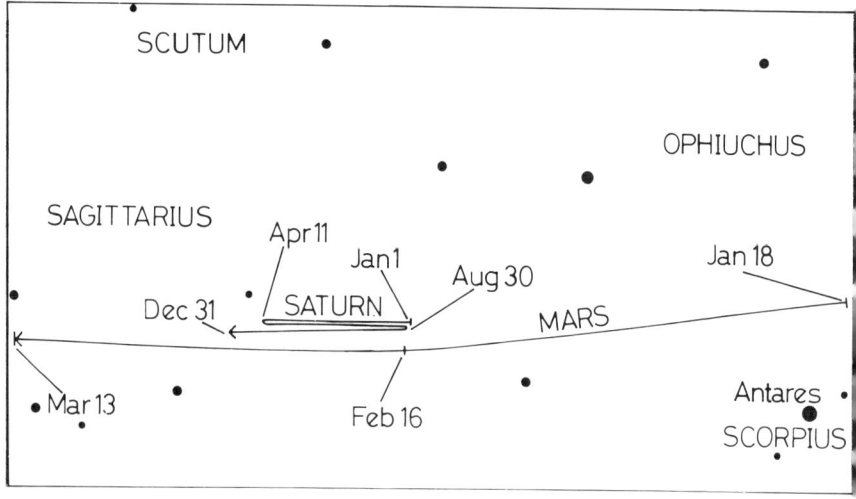

Figure 1. The paths of Mars and Saturn.

MARS AND ANTARES. The Greek name for Mars, the Red Planet, was Ares; the name of Antares, leader of the Scorpion, really means 'Ant-Ares', the Rival of Mars – and certainly the two are of much the same colour. This is strikingly obvious when the two are close together in the sky, as they are during the first weeks of 1988. On the 21st the distance between them will be only 5 degrees – about the same as that between Alpha and Beta Ursæ Majoris (Dubhe and Merak), the Pointers to the Pole Star.

At its brightest Mars can attain magnitude -2.8, and it will almost reach this maximum in September of this year, when it passes through its most favourable opposition for many years. At its faintest, however, it may be little brighter than the Pole Star. During January its magnitude is about 1.5; That of Antares is 1.0, so that Mars is half a magnitude the fainter of the two.

With binoculars, or a small telescope, the two appear not dissimilar at the moment. The apparent diameter of Mars is less than 5 seconds of arc (by September it will have increased to nearly 24 seconds) and so not even large telescopes will show much upon its tiny disk. However, this conjunction presents a good opportunity for amateur photographers, particularly as Saturn also will be available in the same wide-angle photographic field. It is, of course, true that the low altitude will make observations difficult, and the conjunction will not be visible against a really dark background.

THE ASTRONOMICAL ADMIRAL. This month is the centenary of the birth of a famous astronomer – William Henry Smyth, who was born in London on January 21, 1788. He joined the Navy in his youth and fought during the Napoleonic Wars; then, between 1817 and 1824, he carried out a major scientific survey of the Mediterranean. He was always interested in astronomy, and after retiring from the Navy, with the rank of Admiral, he established a major observatory at Bedford, devoting himself to observations of various objects including comets and double stars. He is probably best known for his book *Cycle of Celestial Objects*, containing the Bedford Catalogue of double and multiple stars, clusters, and nebulæ.

Smyth was awarded the Gold Medal of the Royal Astronomical Society, and served as its President for a two-year term. He died in 1865, but in 1879 an English astronomer, Herbert Sadler, published an attack on Smyth's double-star measurements in the *Cycle* claiming that they had been faked. There was an immediate outcry, and the leading American double-star observer, S. W. Burnham, was called in to adjudicate. Before long Burnham found the answer. In some cases, he found Smyth had given estimates for his wide double-star measurements, but, quite innocently, had not made this clear. In fact, Smyth's honesty and reputation were fully vindicated, but it was also true that some of the results for double stars in the *Cycle* were faulty.

Despite this, the *Cycle* still makes fascinating reading, and quite apart from his observational work Smyth deserves to be remembered as one of the great popularizers of astronomy of the last century.

February

Full Moon: February 2   *New Moon:* February 17

Mercury continues to be visible as an evening object, low above the south-western horizon for a very short while at the end of evening civil twilight, though only for the first five or six days of the month. Inferior conjunction occurs on February 11 and for the last ten days of the month it will be visible as a morning object – though not to observers in northern temperate latitudes. Observers nearer the Equator and in southern latitudes should consult the diagram given with the notes for March (Figure 2).

Venus continues to be visible as a brilliant evening object, magnitude −4.1. It is visible in the western sky for several hours after sunset.

Mars, magnitude +1.2, continues to be visible as a morning object as it moves slowly eastwards in Sagittarius. Its reddish tint is a useful aid to identification.

Jupiter continues to be visible as an evening object, in the south-western sky. By the end of February observers in the latitudes of the British Isles will not be able to see the planet after 22h.

Saturn, magnitude +0.6, is a morning object, visible in the south-eastern sky before dawn. The path of Saturn amongst the stars is shown in Figure 1 (given with the notes for January). On February 23 Mars passes 1°.3 S. of Saturn.

THE CELESTIAL WHALE.   Cetus, the Whale, is one of the largest constellations in the sky. It covers an area of 1232 square degrees, and there are only three constellations larger than that (Hydra,

1303 square degrees; Virgo, 1294, and Ursa Major, 1280 – contrast these with the 68 degrees of the smallest constellation, which is, rather surprisingly, Crux Australis, the Southern Cross). Yet Cetus is not particularly striking, and it has only eight stars above the fourth magnitude. They are:

Star	Spec.	Magnitude	Absolute mag.	Distance, light years
β Diphda	K2	2.04	0.2	68
α Menkar	M2	2.53	−0.5	130
η	K2	3.45	−0.1	117
γ	A2	3.47	1.4	75
τ	G8	3.50	5.7	12
ι	K2	3.56	−0.1	163
θ	K0	3.60	0.2	114
ζ Baten Kaitos	K2	3.73	−0.1	189

(The absolute magnitude and distance values are according to the authoritative *Cambridge Sky Catalogue*; in some cases other catalogues give rather different values.)

In mythology, Cetus has been regarded as the fearsome sea monster of the legend of Perseus and Andromeda, though it is often also relegated to the status of a harmless whale. It lies south of Perseus, and is best seen during autumn (northern hemisphere) and spring (southern hemisphere); it intrudes into the ecliptic, so that it may sometimes contain planets, and a small part of it lies north of the celestial equator. The declination of Menkar, the second brightest star in the constellation (though lettered Alpha, it is half a magnitude fainter than Beta or Diphda), is actually just over 4 degrees. The distinctive 'head' of Cetus is made up of Menkar, Mu (magnitude 4.3), Xi (also 4.3) and Delta (4.1). Gamma Ceti has a companion of magnitude 6.4 at a distance of 3 seconds of arc and a position angle of 293 degrees; it appears to be a slow binary. Tau Ceti is one of the nearest known stars, and though smaller and fainter than the Sun it is regarded as a good candidate for the centre of a planetary system; it was one of the two stars studied during the original Ozma experiment of 1960 (the other was Epsilon Eridani).

Undoubtedly the most famous star in Cetus is Omicron or Mira, the long-period variable which has given its name to the entire class of these stars. Its mean period is just over 331 days, and the magnitude ranges between 2 and 10. At some maxima it has even been known to exceed the second magnitude, though in other years it fails to exceed 4. The maximum of February 1987 was bright – the magnitude rose to 2.3, so that for a while Mira was brighter than Menkar – but in general the star is a naked-eye object for only about eighteen weeks in the year. Since the period is roughly eleven months, there are years when maximum occurs with Mira too close to the Sun to be seen. However, 1988 is reasonably favourable, and in January and February there should be no problem in locating the 'Wonderful Star'.

At maximum, Mira seems to have a diameter of 280,000,000 miles (450,000,000 km), larger than that of the Earth's orbit round the Sun. The *Cambridge Sky Catalogue* gives its distance as 96 light-years, though other estimates increase this somewhat. The surface temperature ranges between 2600K and 1900K, but most of the star's energy is in infra-red.

Mira is a double star; the companion is of the 10th magnitude, but the separation is only 0.8 second of arc. There is no doubt that the two components are physically associated, but the revolution period is very long; one estimate gives it as 260 years. The companion appears to be a hot sub-dwarf, and, rather surprisingly, it has been suggested that its mass is actually greater than that of Mira itself.

March

Full Moon: March 3   *New Moon:* March 18

Summer Time in Great Britain and Northern Ireland commences on March 27.

Equinox: March 20

Mercury, for observers in the latitudes of the British Isles, is too close to the Sun for observation throughout March, but for observers further south this is the most favourable morning apparition of the year. Figure 2 shows, for observers in latitude S.35°, the changes in azimuth (true bearing from north through east, south and west) and altitude of Mercury on successive mornings when the Sun is 6° below the horizon. This condition is known as the beginning of morning civil twilight, and in this latitude and at this time of year it occurs about 30 minutes before sunrise. The changes in brightness of the planet are indicated by the relative sizes of the circles marking Mercury's positions at five-day intervals. It will be noticed that Mercury is brightest after it reaches greatest western elongation (27°) on March 8. Thus on February 29 its magnitude is +0.4 while on March 30 it is −0.3.

Venus continues to be visible as an evening object. Its magnitude is −4.2.

Mars is a morning object, magnitude +0.9, in Sagittarius.

Jupiter, magnitude −2.1, is an evening object, visible for several hours in the western sky after sunset. On March 6 Venus passes 2° N. of Jupiter.

Saturn, magnitude +0.5, continues to be visible as a morning object in Sagittarius.

There will be a total eclipse of the Sun on March 17-18; see page 115 of this issue.

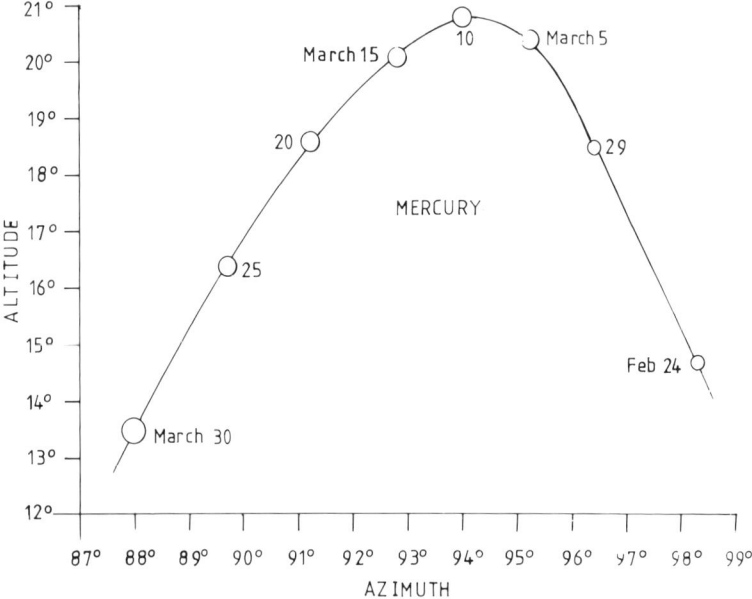

Figure 2. Morning apparition of Mercury for latitude S.35°.

THE SUN'S CORONA. This month's total eclipse of the Sun will not be visible from any part of Europe, but the track covers various suitable areas in the more southerly part of the globe, and though totality lasts for less than four minutes the eclipse will no doubt be well observed. At present the Sun is near the minimum of its cycle of activity (during 1986 and 1987 there were long periods when spots were lacking from the disk), and presumably the corona during the present eclipse will be of the 'minimum' type, though it is always unwise to make predictions.

It is rather surprising to find that not until comparatively modern times was it established that the corona really does belong to the Sun. At the eclipse of April 1567 the astronomer Clavius saw it, but regarded it as merely the uncovered edge of the Sun; Kepler, however, believed the corona to be nothing more nor less than an extensive atmosphere of the Moon. During the 'Maunder Minimum' between 1654 and 1715, when from all accounts sunspots were absent, there is some evidence that during total eclipses the corona was very inconspicuous. In 1806 the Spanish astronomer de Ferrer observed the corona, and pointed out that if it were due to a lunar atmosphere then this atmosphere would have to be fifty times more extensive than that of the Earth, which did not seem reasonable. Final proof came during the eclipses of 1842 and 1851, after which there could no longer be any doubt that the corona was a solar and not a lunar phenomenon.

Certainly this month's eclipse will be worth seeing; details are given in the article by David Allen in this *Yearbook*.

CANOPUS. Canopus, the second brightest star in the sky, is one of Nature's cosmic searchlights. It lies well in the southern hemisphere of the sky, and is never visible from Europe; it can be seen from Alexandria, but not from Athens – one of the early proofs that the Earth is not flat.

In round figures, the declination of Canopus is – 53 degrees. To give the theoretical 'limiting visibility', take 53 away from 90; the answer is 37. Therefore, Canopus should just rise from anywhere south of latitude 37° N. Refraction complicates matters to some extent, and of course it is very hard to see a star which is exactly on the horizon, but Alexandria (latitude 32° N) is quite adequate.

Canopus has a spectrum of type F0, which means that it should be slightly yellowish, but most observers will call it pure white. It is so remote that its distance is to some extent uncertain. The *Cambridge Sky Catalogue* gives the distance as 1200 light-years and the absolute magnitude is -8.5, in which case Canopus is some 200,000 times as luminous as the Sun. Other catalogues reduce these values somewhat, but at least there can be no doubt that Canopus is extremely powerful, and it far outshines Sirius, though from Earth the nearby Sirius appears the brighter by more than half a magnitude. Adopting the Cambridge values, Canopus is well over 7500 times as luminous as Sirius.

Canopus has, incidentally, changed its official designation. In Ptolemy's original list of 48 constellations, much the largest was Argo Navis, the Ship Argo, which in mythology carried Jason and his companions in quest of the Golden Fleece. However, Argo was so large that it was unwieldy, and was finally cut up into various parts. Of these the brightest was Carina, the Keel of the old ship;

and so Canopus, which had been Alpha Argûs, became Alpha Carinæ. Another famous object in this part of the Ship is the variable star Eta, now Eta Carinæ rather than Eta Argûs, which for a time during the last century was the brightest star in the sky apart from Sirius, but is now only on the fringe of naked-eye visibility.

April

Full Moon: April 2   *New Moon:* April 16

Mercury continues to be visible as a morning object for the first week of the month, though not to observers in northern temperate latitudes. Thereafter Mercury is too close to the Sun for observation, passing through superior conjunction on April 20.

Venus continues to be visible as a brilliant evening object, magnitude −4.4, completely dominating the western sky for several hours after sunset. It reaches its greatest eastern elongation (46°) on April 3. For observers in the British Isles Venus will be setting after 23h and towards the end of the month observers in Scotland will have the unusual chance of seeing Venus just above the north-western horizon after midnight!

Mars, magnitude +0.5, continues to be visible in the south-eastern sky in the mornings.

Jupiter, magnitude −2.0, continues to be visible for a short while in the western sky in the early evening, for the first half of the month until it is lost in the evening twilight.

Saturn, magnitude +0.4, continues to be visible as a morning object in Sagittarius. It is better placed for observation by those in southern latitudes where it rises about 3 to 4 hours after sunset but by the end of the month observers in the British Isles should be able to detect it low in the south-east, shortly after midnight.

THUBAN. During April evenings Ursa Major, the Great Bear, known more commonly in America as the Big Dipper, is almost overhead as seen from Britain; from countries such as South Africa and much of Australia it can be seen low over the horizon.

Two of its stars, Merak and Dubhe, show the way to Polaris, the north polar star, which is within one degree of the pole. But Polaris has not always held this distinction; at the time when the Egyptians built their Pyramids, the north polar star was Thuban or Alpha Draconis, in the constellation of the Dragon.

Thuban is easy to find. It lies about midway between Alkaid or Eta Ursæ Majoris, in the Great Bear, and Kocab or Beta Ursæ Minoris, in the Little Bear. However, it is not very brilliant. Its apparent magnitude is 3.65, making it fainter than Megrez, the dimmest of the seven stars of the Plough (3.31). It is a white star of spectral type A0; in the *Cambridge Sky Catalogue* its distance is given as 231 light-years, and it is about 120 times as luminous as the Sun – much less powerful than Polaris.

Thuban is not even the brightest star in the long, winding constellation of Draco; it is exceeded by seven others, of which the leader, Gamma Draconis or Eltamin in the Dragon's head (close to Vega) is of magnitude 2.2. And at present the declination of Thuban is +64½ degrees, so that it is some way from the pole. Yet the phenomenon of precession means that the polar point will eventually return to the neighbourhood of Thuban, so that in the remote future Thuban will again have the distinction of being the north polar star of the sky.

W. T. HAY. This month is the centenary of the birth of one of Britain's most unusual astronomers, W. T. Hay, who was born on December 6, 1888. He became an engineer, but at the age of twenty-one he changed over to a brilliant career in the world of entertainment. He was also interested in flying, and took his pilot's wings. During the First World War he served in the special branch of the RNVR until illness forced him to resign his commission.

By 1932 he had become known as a skilful amateur astronomer, and his observatory at Norbury, in outer London, was equipped with a 12½ inch Calver reflector and a fine 6 inch Cooke refractor. On August 3, 1933 he was making a routine observation of Saturn when he discovered a brilliant white spot in the planet's equatorial region. Nothing of the kind had been seen before, and it caused a tremendous amount of interest when Hay made his announcement; the spot was visible with modest telescopes (the Editor of this *Yearbook*, then aged ten, saw it clearly with a 3-inch refractor). Hay was also a specialist in cometary observations, and devised a special type of cross-bar micrometer. In 1935 he published a small but excellent book, *Through My Telescope*, and he remained active in astronomical work until not long before his death on April 18, 1949.

As an amateur astronomer, then, he is worth remembering; but most people will remember him better as Will Hay, the stage and screen comedian, whose films such as *Oh, Mr Porter!* and *The Ghost of St Michael's* are classics of their kind. Will Hay's particular forte was timing – essential to a comedian, though by no means all comedians possess that elusive gift. In his professional career he was known all over the world, but he always took care to separate it from his very serious and genuine love of astronomy – and he will be remembered by his scientific colleagues for his careful, painstaking work as well as his discovery of the spectacular white spot on Saturn.

May

Full Moon: May 1 and 31   *New Moon:* May 15

Mercury is at greatest eastern elongation (22°) on May 19. It is visible as an evening object throughout the month and for observers in northern temperate latitudes this will be the most favourable evening apparition of the year. Figure 3 shows, for observers in latitude N.52°, the changes in azimuth (true bearing from the north through east, south and west) and altitude of Mercury on successive evenings when the Sun is 6° below the horizon. This condition is known as the end of evening civil twilight, and in this latitude and at this time of year occurs about 40 to 45 minutes after sunset. The changes in brightness of the planet are indicated by the relative sizes of the circles marking Mercury's positions at five-day intervals. It will be noticed that Mercury is brightest before it reaches greatest eastern elongation: on April 29 its magnitude is -1.5 while on May 29 it is $+1.8$. Inexperienced observers may be misled by the appearance of Venus in the same part of the sky, particularly towards the end of the month. Venus is 4 to 5 magnitudes brighter than Mercury.

Venus is still a brilliant evening object, magnitude -4.1, though the period available for observation is decreasing, particularly for observers in the Northern Hemisphere.

Mars is still a morning object in the south-eastern sky, brightening by half a magnitude during the month.

Jupiter passes through conjunction on May 2 and thus unobservable for most of the month. Towards the end of May the planet, magnitude -2.0, becomes visible for a short while low in the eastern sky before dawn.

Saturn continues to be visible as a morning object in Sagittarius. Its magnitude is +0.2.

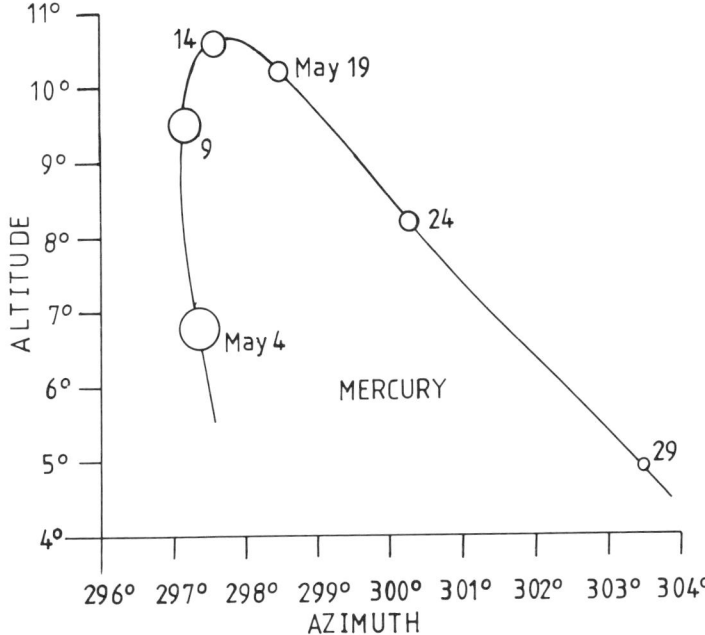

Figure 3. Evening apparition of Mercury for latitude N.52°.

REINMUTH'S COMET. This month sees the return to perihelion of a well-known periodical comet, Reinmuth 1. Unfortunately it will not become bright, but its history is not without interest.

It was originally found on February 22, 1928 by Karl Reinmuth, at Heidelberg Observatory, during a routine search for asteroids. A few pre-discovery plates were subsequently traced. The diameter of the comet's head was about one minute of arc, and the total magnitude was just below 12; there was a short, stumpy tail. It was followed until mid-June, by which time the magnitude had dropped to 17. The period was given by A. E. Levin and J. T. Foxell as 7.2 years, and the comet was recovered on November 5, 1934 very close to the position which Levin and Foxell had given.

Since then the comet has been followed at every return except that of 1942-3, when it was very badly placed. Unfortunately a close approach to Jupiter in 1937 has increased the perihelion distance, and the magnitude does not now rise above 17, so that it is always a very elusive object; no tail has been detected since the return of 1935. Details of its position and magnitude will be given in the 'Comet News' section of the periodical *Astronomy Now*. We may hope that the comet will be recovered fairly early in 1988.

THE ETA AQUARID METEORS, associated with Halley's Comet, are active between May 1 and 8. However, strong moonlight will interfere with the observations of them.

MERCURY AS AN EVENING OBJECT. Northern observers should have no trouble in seeing Mercury with the naked eye during this month. It is, of course, a bright object, but it cannot be seen against a dark background, which makes it elusive. There is a story that the great astronomer Copernicus never saw Mercury in his life, owing to mists rising from the River Vistula in Poland near which he lived, but this seems nothing more than a story of the Canute-and-the-waves type. For one thing, Copernicus spent some time in Italy, where the skies are clear. And when the Editor of this *Yearbook* went to give a lecture in Torún, Copernicus' home town, some years ago he had no problem in locating Mercury with the naked eye, even though pollution and artificial lighting in the area has no doubt increased considerably since Copernicus' time.

However, Mercury is an unrewarding telescopic object, and maps drawn with Earth-based telescopes are bound to be very inaccurate. In 1986 a conference about the planet was held under the auspices of NASA, in Arizona, and your Editor was invited to present a paper on historical maps of Mercury. Even the best of these maps, compiled in the 1920s and 1930s by E. M. Antoniadi with the aid of the powerful 33-inch refractor at the Meudon Observatory in France, bears so litle relation to the truth that the names given cannot be retained; we have lost Antoniadi's 'Solitudo Criophori', 'Solitudo Hermæ Trismegisti' and the rest. Enough that he was much more successful with Mars, where his basic nomenclature has been retained. Virtually all of our knowledge of the Mercurian surface is drawn from the one probe, Mariner 10, which made three active passes of the planet before its power failed.

Even so, we still lack any knowledge of a large part of the surface, and it was admitted at the conference that there is as yet little chance of any new probe. So we must wait until sufficient funds are available; technically it would be easy enough to send another unmanned vehicle to Mercury and complete the mapping, but when it will actually be done remains to be seen.

June

New Moon: June 14   *Full Moon:* June 29

Solstice: June 21

Mercury passes through inferior conjunction on June 13 but by the last week of the month has emerged far enough from the glare of the Sun to be visible as a morning object, though not for observers in northern temperate latitudes. Again it is in the same area of the sky as the much brighter planet Venus, which is slightly further from the Sun.

Venus, magnitude around −3.7, is now moving rapidly towards the Sun and will only be seen for a short while in the western evening sky for the first week of the month. It rapidly passes through inferior conjunction on June 12. On this occasion Venus passes 50′ south of the centre of the Sun. At the next June inferior conjunction the distance is reduced to 30′ while on the next two June occasions (in 2004 and 2012) the distance is so small as to give rise to transits of Venus across the disk of the Sun. For the last ten days of the month Venus may be seen as a morning object low on the eastern horizon before dawn.

Mars, magnitude −0.5, continues to be visible as a morning object. Its path amongst the stars is shown in Figure 4.

Jupiter, magnitude −2.1, is a morning object, visible in the eastern sky before dawn.

Saturn, magnitude 0.0, reaches opposition on June 20 at a distance of 1351 million kilometres. It is in Sagittarius and visible throughout the hours of darkness.

THE NAMES OF SATURN'S SATELLITES.   Saturn is now at opposition.

This is a poor year from the viewpoint of northern observers, since the planet lies in the southernmost part of the Zodiac, but it is excellent for observers in Southern Africa, Australia, New Zealand and South America, since Saturn is very high up and its ring-system is wide open. This is also a good time for observing the satellites. Saturn's satellite family is quite unlike that of Jupiter, which has four large satellites (the Galileans: Io, Europa, Ganymede, and Callisto) and a wealth of small satellites, some of which are retrograde in motion and are probably captured asteroids. Saturn has one large satellite (Titan), half a dozen of intermediate size, and some dwarfs. Their names, given in order of increasing distance from Saturn, are as follows:

> Atlas
> Prometheus
> Pandora
> Janus
> Epimetheus
> Mimas
> Enceladus
> Tethys
> Telesto
> Calypso
> Dione
> Helene
> Rhea
> Titan
> Hyperion
> Iapetus
> Phœbe

Telesto and Calypso are co-orbital with Tethys; Helene is co-orbital with Dione; Janus and Epimetheus periodically 'exchange' orbits, and are almost certainly fragments of an originally single body which was broken up for some reason or other. Of the satellites, only Phœbe has retrograde motion and is presumably asteroidal; moreover Phœbe does not have synchronous rotation (that is to say, it does not keep the same face turned toward Saturn all the time). Unfortunately we do not know a great deal about it, because neither of the Voyager probes passed it closely enough to obtain good pictures.

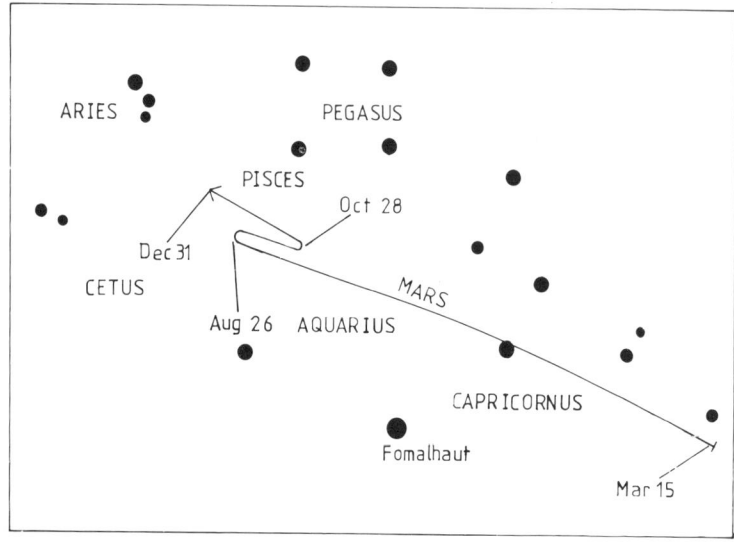

Figure 4. The path of Mars.

Some older books list another satellite moving round the planet in between the orbits of Titan and Hyperion. This satellite was reported by W. H. Pickering in 1904, and was even given a name – Themis – but it has not been recovered, and it is likely that Pickering mistook a star for a satellite. Had Themis existed it would almost certainly have been detected by the Voyagers. On the other hand it is quite likely that some small inner satellites await discovery. We will not know until a new probe is sent to Saturn and its system – which may not be for many years.

FINLAY'S COMET. This periodical comet qualifies as an old friend. It was discovered by W. H. Finlay, from the Cape, in 1886, and since then has been seen at most returns, the latest being that of 1981. At one return (that of 1906) it reached the fringe of naked-eye visibility, though generally it does not develop a perceptible tail. The period is just under seven years.

THE SOUTHERN TRIANGLE. Few constellations give any real impression of the objects after which they are named. One exception is Triangulum Australe, the Southern Triangle, whose three main stars do indeed form a triangle; they are Alpha (magnitude 1.9), Beta (2.8) and Gamma (2.9). Alpha has a K2-type spectrum, and is obviously orange-red in colour. The Triangle lies close to Alpha and Beta Centauri, the Pointers to the Southern Cross, and is therefore too far south to be seen from any part of Europe: the declination of Alpha Trianguli Australe is −60 degrees. There are not many objects of immediate telescopic interest in the group, but the cluster NGC 6025 is easily visible with binoculars.

July

New Moon: July 13   *Full Moon:* July 29

Earth is at aphelion (farthest from the Sun) on July 6 at a distance of 152 million kilometres.

Mercury attains its greatest western elongation (21°) on July 6 and, apart from the last week of the month, is visible as a morning object for observers in equatorial and southern latitudes.

Venus is a morning object, magnitude −4.5, rapidly pulling away from the Sun. By the end of the month Venus is rising about two hours before the Sun.

Mars has now brightened to reach a magnitude of −1.2 by the end of the month when, for observers in the British Isles it becomes visible low above the eastern horizon shortly after midnight.

Jupiter, magnitude −2.2, is now a prominent morning object and by the end of the month is visible by midnight to observers in the British Isles, though for observers in southern latitudes it will still be rising after 02h. Figure 5 shows the path of Jupiter amongst the stars throughout the year.

Saturn, magnitude +0.2, continues to be a prominent object in the night sky though for observers in the latitudes of the British Isles it is never at any great altitude above the horizon since its declination is −22°.

THE R CORONÆ SHELL.   One of the most interesting variable stars in the sky is R Coronæ, now well on view during the evenings. Its declination is 28° N., so that it is reasonably high up from all inhabited countries.

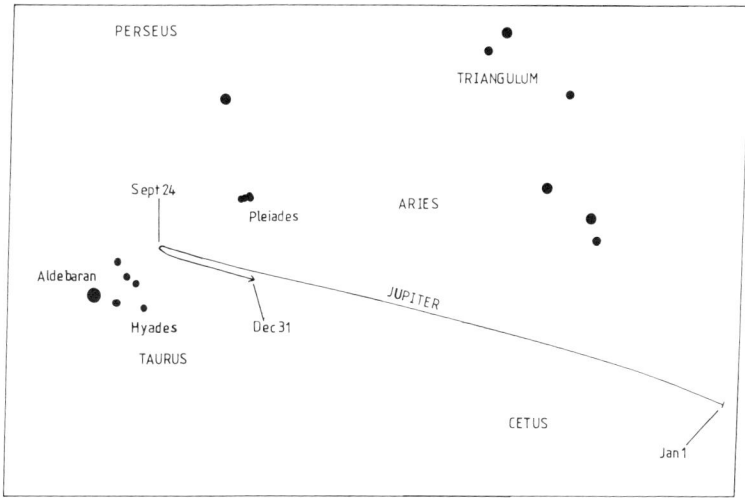

Figure 5. The path of Jupiter.

Corona Borealis (the Northern Crown) is easy to find; it adjoins Boötes, the Herdsman, which is led by the brilliant orange Arcturus. Corona consists of a well-marked semicircle of stars, of which one (Alphekka or Alpha Coronæ) is of the second magnitude. R Coronæ lies in the 'bowl' of the crown. Usually it is on the fringe of naked-eye visibility; binoculars show it easily, together with another star in the 'bowl' whose magnitude is 6.6. Sometimes, however, R Coronæ fades considerably, and may drop at minimum to below magnitude 15, so that powerful telescopes are then needed to show it. These falls are quite unpredictable; after minimum, the star may take weeks or even months to regain its lost lustre.

What happens, apparently, is that R Coronæ is very poor in hydrogen but rich in carbon; sometimes it ejects a cloud of carbon particles which condense into what may only be called soot, masking the star's light until increased radiation from below disperses the soot-clouds. Other stars are known which behave in the same way (SU Tauri is one), but R Coronæ variables are rare, and we cannot yet pretend to understand them completely.

R Coronæ was examined by IRAS, the Infra-Red Astronomical

Satellite which operated for most of 1983, and a remarkable discovery was made. The star is surrounded by a vast dust-cloud, about thirty light-years in diameter – so that if a cloud of similar proportions were centred upon the Sun, stars such as Sirius would be well inside it. According to the infrared results, the mass of the cloud is from 0.3 to 1 per cent of that of the Sun, and seems to have been ejected from the star over a long period ending some 25,000 years ago. But what is heating the cloud? The temperature, around 25K to 30K, seems to be due to some source other than R Coronæ itself, but at the moment nobody is sure what this source may be.

R Coronæ lies in a relatively calm part of the Galaxy, some 4000 light-years above the main plane, which is no doubt why the tenuous shell has been able to persist. Whether other R Coronæ stars have shells of the same kind is not yet known, but strenuous efforts will be made to find out. Certainly the 'sooty star' has presented us with yet another problem.

Because the fades of R Coronæ are so unpredictable, amateur astronomers are encouraged to keep watch and alert professional workers at the first sign of anything unusual. Sometimes the star remains at maximum for years at a time; at other times mimima may follow each other in quick succession. Details of fades are given in periodicals such as *Sky and Telescope* and *Astronomy Now*. So it is always worth looking at the bowl of the Crown through binoculars; if you see only one star instead of two, you may be sure that R Coronæ has started to drop to minimum.

JEAN CHARLES HOUZEAU. One of Belgium's leading astronomers, J. C. Houzeau, died a hundred years ago, on July 12, 1888. He was born at Havre, near Mons, in 1820, and educated at Brussels University; he then erected a private observatory near Mons, and later became assistant at the Brussels Observatory. After a period in America, he returned to the Brussels Observatory as Director in 1876. He produced an authoritative atlas of bright stars and carried out researches into meteoritic astronomy and the Zodiacal Light, but his main claim to fame is probably his great two-volume work on the history of astronomy which was reprinted in England earlier this year.

August

New Moon: August 12   *Full Moon:* August 27

Mercury is at superior conjunction on August 3 and therefore unobservable during the first part of the month. During the second half of August observers in equatorial and southern latitudes will be able to detect the planet as it moves slowly further from the Sun and becomes visible as an evening object. See Figure 6, given with the notes for September.

Venus is a magnificent object in the eastern sky in the mornings reaching its greatest western elongation (46°) on August 22. Its magnitude is -4.3.

Mars is now a prominent object in the sky after midnight, magnitude -1.9. It is moving very slowly in Cetus and reaches its first stationary point on August 26.

Jupiter, magnitude -2.3, continues to be visible as a brilliant morning object. It is in Taurus, moving steadily eastwards between the Pleiades and the Hyades.

Saturn, magnitude $+0.3$, is visible as an evening object. For observers in southern latitudes it is visible until after midnight at the beginning of the month but for those in the latitudes of the British Isles it is too low in the south-western sky for observation by 21h. The rings of Saturn are now open to their widest extent for many years and present a beautiful sight, even in a small telescope. To the naked-eye observer Saturn is brighter when the rings are open since more light is reflected from the Sun. However, in 1988 the effect is slightly diminished since Saturn is at aphelion in September.

GEORG DÖRFFEL AND THE COMET OF 1680. This month's centenary is that of a German amateur astronomer, Georg Samuel Dörffel, who died on August 6, 1688. He was born at Plauen on October 11, 1643, and was educated at Leipzig for the Lutheran ministry; to him astronomy was only a hobby, but a very serious one. He was one of the early observers of the bright comet of 1680, and also computed its orbit.

The 1680 comet was discovered on November 14 by Gottfried Kirch, using a telescope – the first telescopic comet discovery in history. The magnitude was then 4, and there was no tail; Kirch had not been carrying out a deliberate search, but had merely been making some routine early-morning observations of the Moon and Mars. A tail soon developed, and by November 21 the magnitude had risen to 1; Kirch made it equal to Regulus. Perihelion was reached on December 18, and when the comet reappeared in the evening sky the tail was from 70 to 90 degrees long. The brightness then declined, and the last observation was made on March 19, 1681. According to Dörffel, the orbit was essentially parabolic. The comet was observed by both Newton and Halley, though in 1682 the comet we now call Halley's appeared and reached a magnitude of 0 – brighter than the comet of 1680, though it seems that the tail was not so long, and did not exceed 30 degrees.

THE PERSEIDS AND THEIR COMET. The Perseids should be well observed this year; they reach their maximum on August 12, and this is also the day of a new Moon, so that moonlight will not interfere with the best of the Perseids. The usual ZHR is about 70. (The ZHR, or Zenithal Hourly Rate, is the number of naked-eye meteors from the shower which would be expected to be seen by an observer under ideal conditions, with the radiant at the zenith or overhead point; in practice, of course, these conditions never apply, so that the actual observed rate is always less than the ZHR.)

The Perseids are associated with Comet Swift-Tuttle, which was seen for the first and last time in 1862; in September of that year it reached the second magnitude, with a tail almost 30 degrees long. Calculations by F. Hayn (1889) and B. Marsden (1973) gave a period of 120 years, so that the comet should have reappeared in 1984; but up to now there has been no trace of it – so that either the calculated period is wrong, or else the comet has come and gone unseen (it is not likely to have simply disintegrated).

However, Marsden has since drawn attention to the orbital similarity of Swift-Tuttle to that of Kogler's Comet of 1737, though Kogler's Comet was followed only during July and no precise calculations could be made. If, however, the comets really are identical, the period will be about 130 years, and the next return will be in 1992. No doubt searches for it will continue to be made. At any rate, there seems to be no question that Comet Swift-Tuttle really is associated with the reliable Perseid meteor stream.

EQUULEUS. Equuleus, the Foal or Little Horse, covers an area of only 74 square degrees, and is thus the smallest constellation in the sky apart from the Southern Cross. It is well placed during August evenings, and since it lies mainly between 5 and 10 degrees north of the celestial equator it is visible from most parts of the world. In mythology it is said to represent a foal given by Mercury to Castor, one of the Heavenly Twins. There are only three stars above the fifth magnitude: Alpha (3.9), Delta (4.5) and Gamma (4.6), which form a small triangle roughly between Altair in Aquila and Enif in Pegasus. Delta Equulei is a close binary with almost equal components. The orbital period is only 5.7 years, but the pair is difficult to split, since the separation never exceeds 0″.35.

September

New Moon: September 11   *Full Moon:* September 25

Equinox: September 22

Mercury is at greatest eastern elongation (27°) on September 15 and is visible to observers in southern latitudes for whom it is the most suitable evening apparition of the year. Figure 6 shows for observers in latitude S.35°, the changes in azimuth and altitude of Mercury on successive evenings when the Sun is 6° below the horizon. At this time of year this condition, known as the end of evening civil twilight, occurs about 30 minutes after sunset. The changes in the brightness of the planet are roughly indicated by the sizes of the circles which mark its position at five-day intervals. It will be noticed that Mercury is at its brightest before it reaches greatest eastern elongation: on August 27 its magnitude is -0.2 while on September 26 it is $+0.5$.

Venus continues to be a brilliant object, magnitude -4.2, in the eastern morning sky before dawn.

Mars is a brilliant object in the night sky reaching its maximum brightness (magnitude -2.8) towards the end of the month. It is at opposition on September 28, although it is actually closest to the Earth some six days earlier at a distance of 59 million kilometres. This is a very favourable opposition for observers since the planet will not be so close to the Earth again during the rest of this century.

Jupiter, magnitude -2.5, continues to be visible as a morning object in Taurus, reaching its first stationary point on September 24.

Saturn, magnitude $+0.5$, is an evening object in Sagittarius.

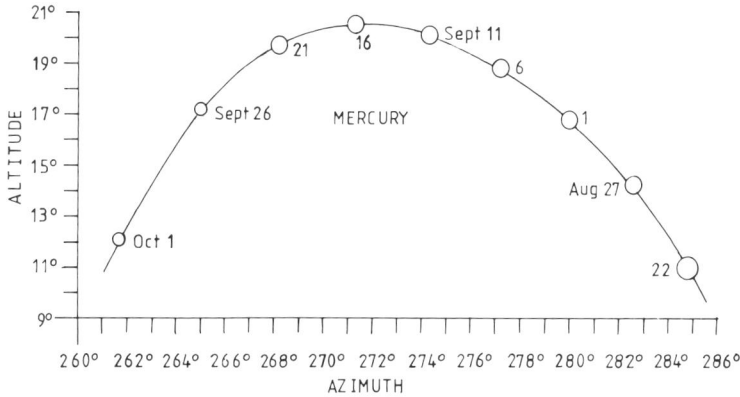

Figure 6. *Evening apparition of Mercury for latitude S.35°.*

RICHARD ANTHONY PROCTOR. There have been various great popularizers of astronomy, ranging from James Ferguson in the early days to Sir Robert Ball in near-modern times and the first astronomical broadcasters such as Sir James Jeans and Sir Arthur Eddington; in America there have been Harlow Shapley, Carl Sagan and many others. In any list of popularizers the name of R. A. Proctor must rank high – so high, indeed, that scientific historians often forget that he was also an excellent cosmologist.

Proctor was born in Chelsea in 1837, and was educated in Cambridge. He intended to follow a legal career, but instead took up astronomy, and made his mark with his book *Saturn and its System* – the first in modern or near-modern times to be devoted to that planet. Proctor was an excellent writer, and other books followed in rapid succession. He was also a skilful observer; his map of Mars in the 1860s was the best available until that of Schiaparelli in 1877. In the field of cosmology, he drew attention to the fact that five of the main stars in the Plough or Dipper have similar motions in space – we now know that they make up what is today termed a moving cluster – and he came to the conclusion that the star-system was not only finite, but also rather smaller than had been generally assumed. In 1881 he emigrated to America, and settled in Missouri, where he died a hundred years ago – on September 12, 1888. His daughter Mary also became a noted astronomer and writer.

THE COMET WITH THE DUSTY TAIL. Comet Tempel 2 comes to perihelion this month. It will not be bright – only at one observed return (that of 1925) has it exceeded the seventh magnitude – but it is of interest because it has been found to have a long dusty tail detectable only in infrared.

The comet was originally found by W. Tempel, from Milan, on July 3, 1873, when its magnitude was 9.5. It was found to have a period of 5.3 years, and it has been seen at most subsequent returns, though sometimes it has been very faint (as in 1978, when the maximum magnitude was only 18.5). Usually it shows no tail, though occasionally it has developed one. In 1983, however, it was examined with equipment on IRAS (the Infra-Red Astronomical Satellite) and a surprising discovery was made. There seemed to be an associated stream of small particles moving through space in parallel paths, and it was established that Tempel 2 does indeed have a long, dusty tail which was quite invisible optically. Other comets examined from IRAS did not show similar tails, and it is too early to say whether or not such phenomena are common. Undoubtedly Tempel 2 will come in for close attention when it returns this year.

THIS MONTH'S LUNAR ECLIPSE. 1988 has not been a good year for eclipses. So far as the Moon is concerned, there is only one (excluding the penumbral eclipse of last March). The eclipse of August 27 will be partial, with less than 0.3 of the Moon's disk covered, and nothing of it will be seen from Europe. However, any lunar eclipses are worth looking at, and they can be quite striking, even though it would be wrong to pretend that they are of much real astronomical importance.

WHERE DOES ALKAID SET? Of the seven famous stars in Ursa Major making up the Great Bear, Plough or Dipper, the southernmost is Alkaid or Benetnasch (Eta Ursæ Majoris). To the nearest degree, its declination is 49° N. Therefore it is circumpolar from anywhere on Earth north of latitude 41° N, and does not rise from anywhere south of latitude 41° S, though of course refraction complicates matters somewhat and in any case it is virtually impossible to see a star which is exactly on the horizon. However, Alkaid is circumpolar from the whole of the British Isles; it sets briefly from New York and Athens; it is doubtful whether it will ever be observable from Wellington in New Zealand, and certainly not from New Zealand's South Island.

October

New Moon: October 10   *Full Moon:* October 25

Summer Time in Great Britain and Northern Ireland ends on October 23.

Mercury is at inferior conjunction on October 11 but then moves rapidly outwards from the Sun to become visible as a morning object during the second half of the month. Greatest western elongation (18°) occurs on October 26. For observers in the Northern Hemisphere this is the most suitable morning apparition of the year. In the latitudes of the British Isles the visibility period is from about October 18 to November 5. Figure 7 shows, for observers in latitude N.52°, the changes in azimuth and altitude of Mercury on successive mornings when the sun is 6° below the horizon. At this time of year and in this latitude this condition, known as the beginning of morning civil twilight, occurs about 35 minutes before sunrise. The changes in the brightness of Mercury are indicated approximately by the sizes of the circles which mark its position at five-day intervals. It will be noticed that Mercury is brightest after it reaches greatest western elongation: on October 26 its magnitude is −0.5 and on November 5 it is −0.8.

Venus, magnitude −4.1, continues to be visible as a brilliant object in the morning sky before dawn.

Mars, magnitude −2.3, is still a brilliant object and being only just past opposition is still visible for the greater part of the night. Mars is moving slowly in Pisces, reaching its second stationary point on October 30.

Jupiter, magnitude −2.7, is a brilliant object and now visible for most of the night. It is moving slowly westwards in Taurus.

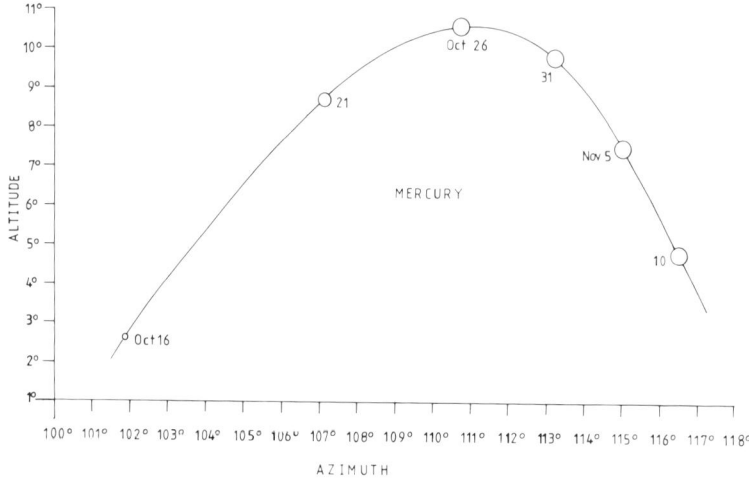

Figure 7. Morning apparition of Mercury for latitude N.52°.

Saturn, magnitude +0.6, continues to be visible as an evening object.

MARS AT ITS BEST. The orbit of Mars is more eccentric than that of the other inner planets; the eccentricity amounts to 0.093 – as against only 0.007 for Venus, whose path is therefore practically circular. The synodic period of Mars is, on average, 780 days, but of course this is not constant; the longest interval between successive oppositions is 810 days, while the shortest interval is only 764 days.

The distance of Mars from the Sun ranges between 249,000,000 km and 206,700,000 km, giving a mean of 227,940,000 km (just over 141 million miles). The most favourable oppositions occur when Mars is near perihelion, as has been the case in 1988, when the minimum distance between the Earth and Mars has been no more than 58,400,000 km (36,300,000 miles). This has been the closest approach since the opposition of 1971 (minimum separation 56,200,000 km).

This year, for once, Northern Hemisphere observers have been favoured, as Mars has been in Pisces. The next few oppositions will not be so close. They are:

Date	Constellation	Max. apparent diameter″	Max. magnitude
1990 Nov.27	Taurus	17.9	−1.7
1993 Jan.7	Gemini	14.9	−1.2
1995 Feb.12	Leo	13.8	−1.0
1997 Mar.17	Virgo	14.2	−1.1
1999 Apr.24	Virgo	16.2	−1.5

This year the apparent diameter has attained almost 24 seconds of arc, and Mars has been the brightest of all the planets apart from Venus; at the oppositions of 1993, 1995 and 1997 it will not even equal Sirius. By then, however our knowledge of the Red Planet should have increased considerably. The Russian probe, to study not only Mars but also its satellites, is due to be launched in the near future; it has been announced that there is to be British collaboration in this project.

The time when serious astronomers still believed in advanced Martian life, and even a canal-building civilization, is long past; up to now the various space-craft have not revealed any trace of organic activity. However, it is still too early to say definitely that Mars is, and always has been, totally sterile. In any case it is much less unwelcoming than any of the other planets in the Solar System apart from Earth, and there is every reason to expect that it will be reached by man during the 21st century – perhaps even before.

ROBERT RECORDE. One of the more interesting of this year's centenaries is that of the death of Robert Recorde, a somewhat shadowy and elusive English astronomer whose career was chequered by any standards. He was born in 1510 of a good family, at Tenby in Pembroke, and went to Oxford in 1525; he acquired a B.A. degree and probably an M.A. as well. He read and taught mathematics and medicine, apparently with great skill. He was awarded an M.D. from Cambridge in 1545, and then went back to Oxford to teach mathematics, rhetoric, anatomy, music, astrology and cosmography. By 1547 he was a practising medical doctor in London, attending both King Edward VI and Queen Mary. In 1556 he published his book *Castle of Knowledge*, which supported the Copernican system, introduced the equality sign (=), and was described as the outstanding astronomical textbook of the sixteenth century. He became Comptroller of the Mint at Bristol, and

Surveyor of the Mines and Monies of Ireland. This, however, marked the climax of his career, and for some reason he was imprisoned, so that during 1588 he died in the King's Bench prison in Southwark. It was a sad end to a highly distinguished career.

November

New Moon: November 9   *Full Moon:* November 23

Mercury is visible as a morning object for the first few days of the month but gradually moves closer to the Sun as it approaches superior conjunction on December 1.

Venus is still visible as a brilliant object in the eastern sky in the mornings before sunrise. Keen sighted observers may detect some change in its brightness since they saw it in July at magnitude -4.5: it is now -4.0.

Mars fades by a whole magnitude during the month, though still a prominent object in Pisces. By the middle of the month it is no longer visible after about 02h.

Jupiter comes to opposition on November 23 at a distance of 603 million kilometres. It is a brilliant object, magnitude -2.9, visible throughout the hours of darkness. Jupiter is moving slowly westwards in Taurus, between the Pleiades and the Hyades. Now is a convenient time for those observers with binoculars to see if they can detect the four Galilean satellites of the planet.

Saturn, magnitude $+0.5$, continues to be visible as an evening object in the south-western sky.

FUTURE MISSIONS TO JUPITER. Jupiter, at opposition this month, is a brilliant object, and a favourite target for the owners of small telescopes; in addition to the constantly changing disk, there are the movements of the four Galilean satellites which have proved to be such fascinating bodies. Ganymede and Callisto are icy and cratered; Europa is icy and smooth, Io red and volcanic. It has been aptly stated that 'there is no such thing as an uninteresting Galilean'.

By now the Galileo probe to Jupiter should have been on its way, following in the wake of the four previous missions – Pioneers 10 and 11, and Voyagers 1 and 2. En route, Galileo was to have obtained the first close-range pictures of an asteroid (Amphitrite). The originally planned launch date was 1982. This was rescheduled for May 20, 1986, from the Space Shuttle *Atlantis*, but then came the *Challenger* tragedy, and the whole American space programme was thrown into confusion.

The new launch date is November 1989, but many aspects of the mission have had to be modified. Galileo will first move inward toward the Sun and fly past Venus at a minimum distance of 19,000 km; a week later it will be at its closest to the Sun (106,000,000 km) and will then move outward again, passing the Earth at 3,600 km. A year later the main rockets will be fired, sending Galileo toward the asteroid belt, but then the probe will move inward again, passing Earth at only 300 km in December 1992. This whole procedure seems complicated – and indeed it is; it has been code-named VEEGA (Venus-Earth-Earth Gravity Assist). Finally, Galileo will make for its main target, and will approach Jupiter in late 1995.

Some five months before the Jupiter rendezvous, Galileo will send out an 118-kg probe which is destined to plunge to destruction in the Jovian clouds; with luck it should maintain contact for an hour or so after entry, relaying its data back by means of the main probe – which will be put into a closed orbit round Jupiter. For the following two years or so Galileo will continue to move round Jupiter, surveying not only the planet but also studying its satellites. At one point it is scheduled to pass within 1000 km of Io, which we already know to be the most volcanically active world in the entire Solar System.

The delay has been unfortunate; the complex VEEGA manœuvres have to be carried out because of the lack of sufficient power; but if all goes well, Galileo will surpass even the results from the Voyagers. We must certainly hope so, because no further American probes to the outer part of the Solar System have been funded as yet, and there is no definite news of any Russian attempts.

G. D. MARALDI. November 14, 1788, marked the death of a

well-known Italian astronomer, Giovanni Domenico Maraldi. He was born at Perinaldo in 1709; his father, G. F. Maraldi, was also an expert astronomer and was particularly noted for his observations of the planets (he was the nephew of G. D. Cassini, first Director of the Paris Observatory, after whom the main division in Saturn's ring-system is named). G. D. Maraldi also spent most of his career at the Paris Observatory, and made careful, long-continued observations of the positions of the four main satellites of Jupiter.

December

New Moon: December 9   *Full Moon:* December 23

Solstice: December 21

Mercury is at superior conjunction on December 1 and then moves slowly eastwards from the Sun. It remains unsuitably placed for observation throughout the month except for observers in equatorial latitudes who may be able to glimpse it as an evening object for the last few days of the month.

Venus, magnitude -4.0, continues to be visible as a brilliant object in the eastern sky in the mornings though the period available for observation is shortening noticeably as it moves towards the Sun. By the end of the year Venus will not rise until after 07h, for observers in southern Britain, while in the north of Scotland it will be nearly an hour later.

Mars, magnitude -0.5, continues to be visible in the western sky until after midnight.

Jupiter, magnitude -2.8, is just past opposition and thus visible as a brilliant object in Taurus for the greater part of the night.

Saturn, magnitude $+0.5$, is only visible for a short while in the evenings, low in the south-western sky. Observers are unlikely to see it after the first ten days of the month.

AN ELUSIVE COMET. The periodical Comet du Toit 1 is due to return to perihelion this month, reaching its closest point to the Sun on Christmas Day. It is likely to be very faint, and at one time it was thought that it would probably be lost.

It was discovered on May 16, 1944 by D. du Toit at the Boyden

Observatory at Bloemfontein, in South Africa; it was then of the 10th magnitude, but it faded quickly, and was lost by the end of November. The period was calculated by F. J. Bobone as being 14.8 years, which would mean a return in 1959; however, careful searches made at Flagstaff by Elizabeth Roemer, one of the leading comet-hunters, failed to show it. The following return was due in 1974; searches were made by Miss Roemer and also by Charles Kowal (Palomar) and C. Torres (Cerro el Roble), but again the results were negative, and Comet du Toit was more or less given up. However, in January 1975 Torres was re-examining his plates when he detected a diffuse image of magnitude 19 which looked as though it might be the comet; it was also found on a second plate taken on the previous April 22. Apparently the comet had been too faint to be recorded on Kowal's photographs and was outside the region covered by Roemer's.

Whether the comet will reappear this year remains to be seen, but there seems no reason why it should be missed provided that intensive searches are made for it.

NEARBY BRIGHT STARS. Eridanus, the River, is well placed during evenings in December. Its leading star, Achernar, is in the far south of the sky; but also of interest is Epsilon Eridani, which is one of the two nearest stars to be not too dissimilar from the Sun (the other is Tau Ceti). Epsilon Eridani is easily visible with the naked eye, and as a matter of interest below is a list of other stars with apparent magnitude 4.5 or brighter and whose distances are 40 light-years or less.

By no means all of these are suitable as candidates for the centres of planetary systems – indeed, some of them could hardly be less promising. However, they are at least in our own part of the Galaxy, and one never knows!

Stars with apparent magnitude 4.5 or brighter and with distances of 40 light-years or less

Star	Apparent mag.	Spectrum	Luminosity $Sun=1$	Distance light years	
Beta Aquilæ	3.7	G8	3.2	36	Alshain
Alpha Boötis	−0.04	K2	115	36	Arcturus

Xi Boötis	4.5	G8	0.8	22
Beta Canum Venaticorum	4.3	G0	1.3	30
Alpha Canis Majoris	−1.5	A1	26	8.7 Sirius
Alpha Canis Minoris	0.4	F5	11	11.4 Procyon
Phi Capricorni	4.1	F5	3	39
Eta Cassiopeiæ	3.4	G0	5.9	19
Alpha Centauri	−0.3	K2,G1	1.3	4.3
Tau Ceti	3.5	G8	0.4	11.9
Beta Comæ Berenices	4.3	G0	1.2	27
Gamma Coronæ Australis	4.2	F8	2.1	39
Chi Draconis	3.6	F7	2	25
Sigma Draconis	4.7	K0	0.3	18.5
e Eridani	4.3	G5	0.4	20
Epsilon Eridani	3.7	K2	0.25	10.7
Delta Eridani	3.5	K0	2.3	29 Rana
Omicron^2 Eridani	4.4	K1	0.3	16 Keid
Beta Geminorum	1.1	K0	60	36 Pollux
Zeta Herculis	2.8	G0	5.2	31 Rutilicus
Mu Herculis	3.4	G5	2	26
Beta Hydri	2.8	G1	7	21
Alpha Hydri	2.9	F0	8	36
Epsilon Indi	4.7	K5	0.17	11.2
Beta Leonis	2.1	A3	18	39 Denebola
Gamma Leporis	3.6	F6	2	26
Epsilon Muscæ	4.1	gM6	11	39
36 Ophiuchi	4.3	K0	0.2	17.6
70 Ophiuchi	4.0	K0	0.4	16.6
Pi^3 Orionis	3.2	F6	3	25
Chi^1 Orionis	4.4	G0	1.2	32
Delta Pavonis	3.6	G5	1	18.6
Iota Persei	4.0	G0	2.6	39
Alpha Piscis Australis	1.2	A3	22	22 Fomalhaut
Omega Sagittarii	4.7	dG5	0.7	36
Delta Sculptoris	4.5	A0	50	30
Lambda Serpentis	4.4	G0	1.2	36
Zeta Tucanæ	4.2	G0	0.8	23
Xi Ursæ Majoris	3.8	G0	0.9	25
Beta Virginis	3.6	F8	2.7	33 Zavijava
Gamma Virginis	2.7	F0+F0	8	36 Arich

Eclipses in 1988

1. Penumbral eclipses of the Moon

These are never given in these eclipse notes, since they only dim the brightness of the Moon and are often unobservable. It is now known that the lunar eclipse of March 3, which was listed in the Astronomical and Calendarial Data Sheet for 1988 and in Planetary and Lunar Coordinates 1984-2000 as a partial eclipse of the Moon of magnitude 0.0022, will be a penumbral eclipse. The time of maximum eclipse is $16^h\ 13^m$.

2. A total eclipse of the Sun on March 17-18

The path of totality begins in the Indian Ocean to the west of Sumatra, crosses Sumatra, Borneo, and the southern part of the Philippine Islands, passes close to the Mariana Islands and south of the Aleutian Islands and ends in the Gulf of Alaska. The partial phase is visible from eastern Asia, Indonesia, north-western Australia, New Guinea, Micronesia, the extreme north-west of North America and the western Hawaiian Islands. The eclipse begins on March 17 at $23^h\ 24^m$ and ends on March 18 at $04^h\ 32^m$; the total phase begins on March 18 at $00^h\ 23^m$ and ends at $03^h\ 32^m$. The maximum duration of totality is $3^m\ 46^s$.

3. A partial eclipse of the Moon on August 27

This will be visible from North America except the eastern part, Central America, the western part of South America, part of Antarctica, the Pacific Ocean, Australasia and the eastern part of Asia. The eclipse begins at $10^h\ 08^m$ and ends at $12^h\ 00^m$. The time of maximum eclipse is $11^h\ 04^m$, when 0.29 of the Moon's diameter is obscured.

4. An annular eclipse of the Sun on September 11

This event is visible as a partial eclipse from the eastern part of

Africa except for most of Egypt and western Sudan, the southern part of Asia, the Indian and Southern Oceans, Indonesia, Australia except for the extreme north-east, New Zealand and part of Antarctica. The eclipse begins at $01^h 46^m$ and ends at $07^h 41^m$. The annular phase begins at the coast of Somalia at $02^h 59^m$, crosses the Indian and Southern Oceans and ends at $06^h 28^m$ between New Zealand and Antarctica. The maximum duration of the annular phase is $6^m 52^s$.

Occultations in 1988

In the course of its journey round the sky each month, the Moon passes in front of all the stars in its path and the timing of these occultations is useful in fixing the position and motion of the Moon. The Moon's orbit is tilted at more than five degrees to the ecliptic, but it is not fixed in space. It twists steadily westwards at a rate of about twenty degrees a year, a complete revolution taking 18.6 years, during which time all the stars that lie within about six and a half degrees of the ecliptic will be occulted. The occultations of any one star continue month after month until the Moon's path has twisted away from the star but only a few of these occultations will be visible at any one place in hours of darkness.

There are six occultations of planets in 1988, two of Mercury, three of Venus and one of Mars. None of these is visible from Great Britain or North America.

Only four first-magnitude stars are near enough to the ecliptic to be occulted by the Moon; these are Regulus, Aldebaran, Spica, and Antares. Of these four all except Aldebaran are occulted on several occasions in 1988.

Predictions of these occultations are made on a world-wide basis for all stars down to magnitude 7.5, and sometimes even fainter. Lunar occultations of radio sources are also of interest – remember the first quasar, 3C.273, was discovered as the result of an occultation.

Recently occultations of stars by planets (including minor planets) and satellites have aroused considerable attention.

The exact timing of such events gives valuable information about positions, sizes, orbits, atmospheres and sometimes of the presence of satellites. The discovery of the rings of Uranus in 1977 was the unexpected result of the observations made of a predicted occultation of a faint star by Uranus. The duration of an occultation by a satellite or minor planet is quite small (usually of the order of a minute or less). If observations are made from a number of stations it is possible to deduce the size of the planet.

The observations need to be made either photoelectrically or visually. The high accuracy of the method can readily be appreciated when one realizes that even a stop-watch timing accurate to $0^s.1$ is, on average, equivalent to an accuracy of about 1 kilometre in the chord measured across the minor planet.

Comets in 1988

The appearance of a bright comet is a rare event which can never be predicted in advance, because this class of object travels round the Sun in an enormous orbit with a period which may well be many thousands of years. There are therefore no records of the previous appearances of these bodies, and we are unable to follow their wanderings through space.

Comets of short period, on the other hand, return at regular intervals, and attract a good deal of attention from astronomers. Unfortunately they are all faint objects, and are recovered and followed by photographic methods using large telescopes. Most of these short-period comets travel in orbits of small inclination which reach out to the orbit of Jupiter, and it is this planet which is mainly responsible for the severe perturbations which many of these comets undergo. Unlike the planets, comets may be seen in any part of the sky, but since their distances from the Earth are similar to those of the planets their apparent movements in the sky are also somewhat similar, and some of them may be followed for long periods of time.

The following periodic comets are expected to return to perihelion in 1988:

Comet	Year of discovery	Period (years)	Predicted date of perihelion 1988
Reinmuth 1	1928	7.3	May 9
Finlay	1886	7.0	June 6
Tempel 2	1873	5.3	Sept. 16
Longmore	1975	7.0	Oct. 12
du Toit 1	1944	14.7	Dec. 25

Minor Planets in 1988

Although many thousands of minor planets (asteroids) are known to exist, only 3,000 of these have well-determined orbits and are listed in the catalogues. Most of these orbits lie entirely between the orbits of Mars and Jupiter. All of these bodies are quite small, and even the largest, Ceres, is believed to be only about 1,000 kilometres in diameter. Thus, they are necessarily faint objects, and although a number of them are within the reach of a small telescope few of them ever reach any considerable brightness. The first four that were discovered are named Ceres, Pallas, Juno and Vesta. Actually the largest four minor planets are Ceres, Pallas, Vesta and Hygiea. Vesta can occasionally be seen with the naked eye and this is most likely to occur when an opposition occurs near June, since Vesta would then be at perihelion. In 1988 Ceres will be at opposition on September 17 (magnitude 7.4), Pallas on August 2 (magnitude 9.5) and Vesta on January 22 (magnitude 6.6).

A vigorous campaign for observing the occultations of stars by the minor planets has produced improved values for the dimensions of some of them, as well as the suggestion that some of these planets may be accompanied by satellites. Many of these observations have been made photoelectrically. However, amateur observers have found renewed interest in the minor planets since it has been shown that their visual timings of an occultation of a star by a minor planet are accurate enough to lead to reliable determinations of diameter (see page 118). As a consequence many groups of observers all over the world are now organizing themselves for expeditions should the predicted track of such an occultation cross their country.

In 1984 the British Astronomical Association formed a special Minor Planet Section.

Meteors in 1988

Meteors ('shooting stars') may be seen on any clear moonless night, but on certain nights of the year their number increases noticeably. This occurs when the Earth chances to intersect a concentration of meteoric dust moving in an orbit around the Sun. If the dust is well spread out in space, the resulting shower of meteors may last for several days. The word 'shower' must not be misinterpreted – only on very rare occasions have the meteors been so numerous as to resemble snowflakes falling.

If the meteor tracks are marked on a star map and traced backwards, a number of them will be found to intersect in a point (or a small area of the sky) which marks the radiant of the shower. This gives the direction from which the meteors have come.

The following table gives some of the more easily observed showers with their radiants; interference by moonlight is shown by the letter M.

Limiting dates	Shower	Maximum	R.A.	Dec.	
Jan. 1-6	Quadrantids	Jan. 4	15^h28^m	$+50°$	M
April 20-22	Lyrids	April. 21	18^h08^m	$+32°$	
May 1-8	Aquarids	May. 5	22^h20^m	$+00°$	M
June 17-26	Ophiuchids	June 19	17^h20^m	$-20°$	
July 15-Aug.15	Delta Aquarids	July 27	22^h36^m	$-17°$	M
July 15-Aug.20	Pisces Australids	July 31	22^h40^m	$-30°$	M
July 15-Aug.25	Capricornids	Aug. 2	20^h36^m	$-10°$	M
July 27-Aug.17	Perseids	Aug. 12	3^h04^m	$+58°$	
Oct. 15-25	Orionids	Oct. 22	6^h24^m	$+15°$	M
Oct. 26-Nov.16	Taurids	Nov. 3	3^h44^m	$+14°$	M
Nov. 15-19	Leonids	Nov. 17	10^h08^m	$+22°$	
Dec. 9-14	Geminids	Dec. 13	7^h28^m	$+32°$	
Dec. 17-24	Ursids	Dec. 22	14^h28^m	$+78°$	M

M=moonlight interferes

Some Events in 1989

ECLIPSES

There will be four eclipses, two of the Sun and two of the Moon.

February 20: total eclipse of the Moon – N. America, Australasia, Asia, E. Africa, N. E. Europe.
March 7: partial eclipse of the Sun – Hawaii, N. America, Greenland, Asia.
August 17: total eclipse of the Moon – Asia, Europe, Africa, the Americas.
August 31: partial eclipse of the Sun – S. E. Africa.

THE PLANETS

Mercury may be seen more easily from northern latitudes in the evenings about the time of greatest eastern elongation (May 1) and in the mornings around greatest western elongation (October 10). In the Southern Hemisphere the corresponding dates are February 18 (morning) and August 29 (evening).

Venus is visible in the mornings until March, and in the evenings from May to December.

Mars is visible in the evenings until July and in the mornings from November onwards.

Jupiter is at opposition on December 27.

Saturn is at opposition on July 2.

Uranus is at opposition on June 24.

Neptune is at opposition on July 2.

Pluto is at opposition on May 4.

PART TWO

Article Section

PAUL MURDIN and ALEC BOKSENBERG

The William Herschel Telescope

The 4.2-metre William Herschel Telescope (WHT) on La Palma in the Canary Islands began operating in 1987. The telescope completed the original plan for three astrophysical telescopes built at the Roque de los Muchachos Observatory by the Royal Greenwich Observatory (RGO) for astronomers in Britain, the Netherlands, and Spain. The idea of the telescope was to produce the largest possible telescope with the finest possible technical specification within the money available. The excellence of the astronomical conditions on La Palma dictated stringent limits for the sharpness of its images and the accuracy with which it tracked stars. The most important of these from the point of view of designing the telescope was the fine seeing, with star images of 1.0 second of arc or better 40 per cent of the time; at their best star images are 0.3 to 0.6 second of arc.

1987 was the end of a long project, after years of calculations and design work at RGO, and of testing and construction in the Grubb Parsons factory. Under the blunt and outspoken, skilled and energetic project engineer Brian Mack, the William Herschel Telescope was reconstructed in the dome at an altitude of 2400 metres on the Roque de los Muchachos. It is the world's third largest telescope with a single main mirror. We expect that the already proven excellent observing conditions on La Palma, together with the most up-to-date instruments and detectors, are giving it the edge over its larger rivals, the 5-metre telescope in the USA and the 6-metre telescope in the USSR. For a time during its erection it was possible to push the 160-ton telescope around by hand, but now the gears have been meshed and it can be accurately controlled. It towers above the visitor who enters the dome at ground level. The dome is small and hugs the telescope – no voluminous spaces to be filled with hot air that causes convec-

Figure 1. The Dome of the William Herschel Telescope.

tion and therefore bad seeing. The dome is cold – not just because it is unheated but because of the screens and cavity walls that protect the telescope from the warmth of sunlight striking the dome and its supporting wall.

Altazimuth design

The telescope design is based on construction principles by James Nasmyth, a Victorian engineer, inventor of the steam-hammer and holder of numerous patents in canal making. A keen amateur astronomer, Nasmyth built the first 'trunnion-vision' telescope on an altazimuth mounting.

Altazimuth mountings are like guns, with the altitude motion of the telescope corresponding to elevation, and the azimuth corresponding to the bearing around the horizon; they were, of course, the commonest way of holding a telescope before the invention of the equatorial mounting. Nasmyth's idea was to add a third mirror to a Cassegrain telescope mounted in an altazimuth mount, such that the light beam was deflected out of the telescope tube along the altitude axis, through the trunnion which supported the altitude bearing. Nasmyth could sit on a stool mounted on the azimuth bearing with his eye at the trunnion, in relative comfort. He was too old to climb a ladder to the stars, he said, so he brought the stars down to him. The deflected foci are called the Nasmyth foci. Nasmyth's telescope was of such quality that he was the first to see the solar granulation, described by him as a 'willow leaf pattern'.

The altazimuth-mounted William Herschel Telescope has two Nasmyth foci deflected by the third flat mirror (called the Nasmyth mirror), as well as the conventional Cassegrain focus. At the Nasmyth foci are two large horizontal platforms, which can take large instruments, holding them horizontal, fixed relative to gravity.

Nasmyth instruments

By contrast to the rigidly boxed instruments mounted at the Cassegrain focus, which moves to all altitudes, instruments at the Nasmyth foci can be simple, optical benches as if in a laboratory. This gives astronomers the opportunity to try out novel ideas in an instrument in an experimental way, without having to constrain their imagination as well as the components by enclosing the instrument in a rigid box.

The Nasmyth mirror gives astronomers who use the Herschel Telescope another advantage – it is easily switchable to direct the light beam to either of the Nasmyth foci, or to fold out of the way so that the light beam can reach the Cassegrain focus. Thus an astronomer can have at least three instruments on standby through the night to deploy on his programme as it is necessary; alternatively he can switch from one instrument to another as conditions change through the night.

To give an example of both these scenarios:

Astronomer A has the position of an EXOSAT X-ray source observed by the satellite. She uses the low-resolution spectrograph at the Cassegrain focus to survey the optical objects which might be responsible for the X-rays. On the real-time display of data from one of these she sees that it has an emission line spectrum, indicating the presence of gas which might be being accreted on to a compact object such as a neutron star or blackhole. She switches to the high resolution spectrograph for a closer, longer look at the emission lines. She can see the λ4686 ionized helium emission line which is enhanced by a flux of X-rays, so she is pretty sure this is the X-ray candidate, and she notices that H-beta is double, indicating that the gas is flowing round an accretion disk, one side receding, the other approaching. She takes a second exposure. On the second spectrum she sees that the ratio of the two components has altered, so the orbital period might be small. She alternates for the rest of the night between exposing spectra and CCD images, the latter giving her the light curve of the binary star, the former its velocity curve.

Astronomers X, Y and *Z*, on the other hand have quite distinct requirements. *X* has a new infrared camera set up on the Nasmyth focus. She uses it in the afternoon to study star-formation regions; the sky is as black in the daytime as at night to an infrared astronomer so she isn't worried by working in the daytime. At twilight she hands the telescope over to *Astronomer Y* who needs high dispersion spectra of bright stars. It does not matter to him that there is a moderate background of sun- or moon-light. *Y* switches the Nasmyth mirror to the other Nasmyth focus, where there is a high dispersion spectrograph, and follows his programme in twilight and while the moon is above the horizon, but when the Moon sets he hands over to *Z*. Now *Z* folds the Nasmyth mirror away and uses the telescope at the Cassegrain focus while the sky is dark to obtain low dispersion spectra faint galaxies, to obtain redshifts of distant clusters: their light would be swamped by moonlight.

Each of these three astronomers gets the conditions which she can use, and none of them is inconvenienced by adverse conditions which would affect her programme. Scheduling of the telescope can be flexible, without being constrained by the usual night-by-night division of telescope time.

Telescope drives and control

Because the WHT is symmetric relative to gravity, and compact, it is very responsive to its controls. Essentially because the lengths of its structures are small their frequencies of oscillation are high, and

THE WILLIAM HERSCHEL TELESCOPE

Figure 2. The altazimuth-mounted telescope

they respond positively to low frequency pushes. For this reason, the altazimuth mount is liked by engineers. It is also economical since the dome it sits in has no wasted space, hugging the symmetric telescope closer than an off-centre equatorial mounting.

The altazimuth mounting has some special features that optical astronomers are not used to, although radio astronomers will find them familar when they come to observe. The WHT can rotate by a maximum of 0-95° from the horizontal and ±270° about East and astronomers will have to be careful in slewing to their next object always to 'unwind' the azimuth rotation, otherwise they may be stopped during the next integration when they reach a 270° limit. The maximum slewing speed for both axes is 1° per second, with acceleration during slewing reaching a maximum of 0.3° per second per second. These limits set the size of the 'blind spot' at the zenith; the telescope cannot rotate in azimuth by 180° instantaneously if the object happens to pass through the zenith. On the rare occasions when this happens, observing will have to cease temporarily. For the WHT the blind spot is 0.2 degrees radius and the interruption lasts at most three minutes.

In normal use, all the telescope drives are controlled from an operations desk by means of a computer system. All the instrument turntables and cable wrap devices are controlled in sympathy with the telescope motion, as well as the positions of the dome observing slit and windscreen.

Turntables play a bigger role in altazimuth telescopes like the WHT than in equatorial telescopes. Imagine the telescope tracking a nebula up to the zenith. The preceding side of the image is upwards. After passing the zenith, it is the following side of the image which is upwards. If an instrument is fixed to the telescope and remains stationary relative to it, then before the zenith its 'upward' side receives the western portion of the image, afterwards the eastern. The image rotates relative to the instrument and, to follow it, the instrument should itself rotate. This is the philosophy adopted at the Cassegrain focus – the instrument is mounted on a turntable to follow the image. At the Nasmyth foci, however, an optical system of reflections is used to de-rotate the image and to keep it stationary on a fixed instrument.

Telescope optics

So as not to degrade the excellent La Palma seeing, the WHT has the finest large optics money can buy. The telescope is of classical Cassegrain optical configuration. The paraboloidal primary mirror is made of a glass-ceramic material (Cervit) having near-zero coefficient of expansion over the operating temperature range. As the temperature of the mirror falls during the night, the mirror contracts completely imperceptibly and retains its expensively bought figure. The mirror has a clear aperture of 4.2 metres and a focal length of 10.5 metres (f/2.5). The precise diameter of 4.2 metres was determined by the availability of the mirror blank, made by Owens-Illinois as part of the 4-metre series of blanks for telescopes of the 1970s. The mirror, figured by David Brown of Grubb Parsons, concentrates 85 per cent of the light of a distant star into an area only 0.3 second of arc in diameter.

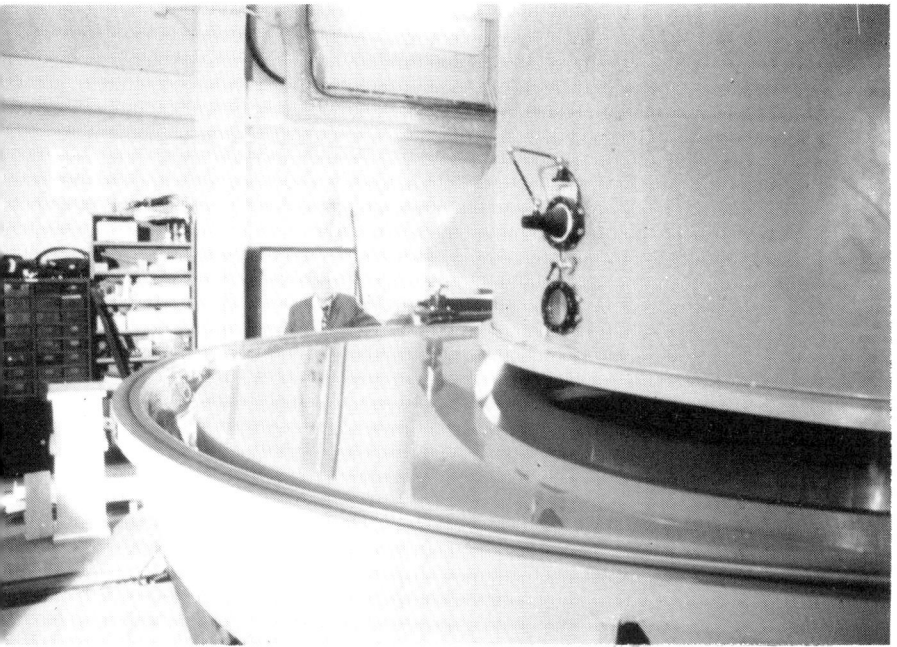

Figure 3. The main mirror being taken out of the aluminizing tank. Your Editor is on the far side to provide comparison.

The optical performance of a telescope depends on controlling the deformation of the mirror surface when the mirror is contained in a mirror cell. In fact, a large mirror would bend by hundreds or thousands of times the optical tolerance if not mounted properly. The images would be useless. The problems for large telescopes become severe very rapidly, since the deflections of a structure increase with its weight multiplied by the lever arm at which the weight is applied, and divided by the cross-sectional area of its material, which provides the stiffness. The deflections of a telescope thus increase as the square of the telescope mirror size, so that in this sense a 4.2-metre telescope is about three times more difficult to make than a 2.5-metre telescope.

The WHT's mirror's diameter-to-thickness ratio of 8 makes it thinner than for most large telescopes built in recent years, but it is not really classifiable as a thin mirror and raises no unusual problems for its support system. Calculations for the WHT mirror show that when it stands on edge in its cell pointing to an object on the horizon, a ridge or wrinkle appears at the lower edge, as the mirror compresses under its own weight; however, the height of the wrinkle is only $^1/_{10}$ the wavelength of light and does not significantly disturb the figure of the mirror.

Dome and building

The telescope is supported by a reinforced concrete pier mounted on piles driven through the volcanic ash to a thick layer of basalt rock 20 metres below the surface. The pier puts the centre of rotation of the telescope at a height of 13.4 metres above the ground. The dome is 21 metres internal diameter, and a pair of up-and-over shutters with a windscreen coupled to the lower shutter will allow observations down to 12° above the horizon. The dome is onion-shaped to allow the up-and-over shutter to track back past the zenith. Sideways moving bi-parting shutters are not used because they stick out from the dome like sails and not only spoil the flow of air over the dome but sustain enormous forces in high winds. The dome is capable of supporting its own weight in ice build-up. The dome is supported on a rail set on to a cylindrical concrete building structure. Set on one side of the cylindrical drum is a three-story rectangular annex. This contains the mirror aluminizing plant, the operations control room, computer room, dark rooms, workshops, offices and various services. Because no unnecessary activity takes place in the dome there is very little thermal disturbance of the air near the telescope, which greatly improves the chance of achieving perfect 'dome seeing'.

Remote observing and future instruments

The WHT will be used at first with standalone instruments, but in the near future RGO and its university partners expect to create for it a versatile suite of interconnected instruments. The switchability of the Nasmyth mirror and the opportunity to reconfigure the telescope quickly means that instruments must be kept on standby, connected to the instrument control computer. Their standby heat is removed from the telescope by a refrigerated cooling system. Each instrument will be controlled locally by its own microprocessor connected by a utility network called Ether-

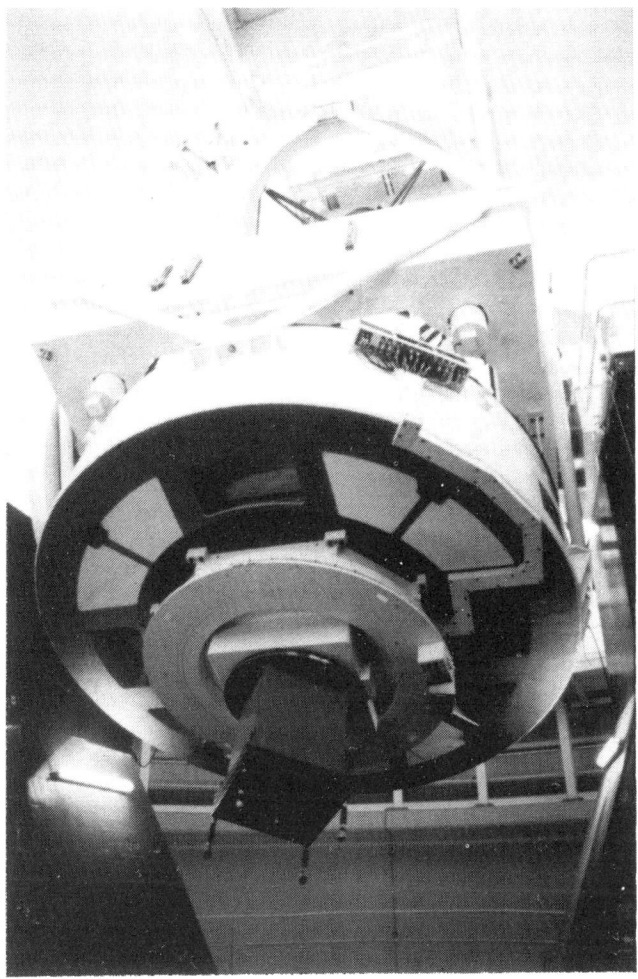

Figure 4. Another view which emphasizes the massiveness of the mountings.

net. The instruments can be commanded and data collected via this network. The VAX computer which is in overall charge of the instruments can host STARLINK-type software for data reduction and can be connected via intercontinental packet-switch networks to computers anywhere in the world. In this way astronomers will be able to observe on the WHT without leaving their home desks. This will be convenient and even efficient; but the real effectiveness of any telescope depends hugely on the capability of the instruments it feeds. Efficient instruments mean that as much as possible of the available light can be used. Good seeing, good optics and good tracking mean that unwanted light from the sky can be excluded from the instruments and make the measurements clearer. The lack of artificial light in the La Palma sky (if lighting development can continue to be controlled) means that the unwanted light is, in any case, small. All this adds up to good prospects for making measurements of fainter stars, galaxies, and quasars. And fainter means, usually, more distant – the William Herschel Telescope is designed for cosmological problems, and named fittingly after Britain's greatest and first observational cosmologist.

These are, then the prospects for the future. The present reality is that British, Dutch and Spanish astronomers are travelling to the Canary Islands, and the La Palma mountain top, to use what we expect is at this time proving to be one of the finest telescopes in one of the finest sites in the world.

DAVID ALLEN

The Shadow Chasers

In March this year the Moon's shadow will etch a charcoal line across the Indian and Pacific Oceans, and through parts of the East Indies. Those fortunate enough to stand in the track of that shadow will witness one of the most impressive and spectacular of phenomena we can experience. Thousands will journey to the shadow zone to absorb the spectacle. Cruise ships will take some into the China Sea, while others will fly to the Philippine island of Mindanao. For many, the eclipse expedition will be their principal holiday of 1988. Why this interest?

That question is one that can be asked only by those who have not been witness to a total eclipse of the Sun. It is all too easy, for the uninitiated, to suppose that these events are just another of the odd things in the sky that amateur astronomers deem entertaining but that ordinary mortals find quaintly boring. Remember all that fuss about Halley's pathetic comet?

Yet rare indeed is the man or woman who can view impassionately the event. It is not merely a phenomenon to add to one's bag of oddities, much as the stamp or butterfly collector can go into raptures over a drab but rare specimen. It is an event of exquisite and undeniable beauty; but not just that either. Most of all, and most impressively, it is an emotional experience, a phenomenon that rips into one's subconscious. Eerie happenings crowd together into scant seconds as totality commences, and everything screams at us that the world has gone awry. As we grapple with an emotional high we are confronted by the vista of beauty as corona and prominences burst into view. Though words subsequently fail and memories grow dim, no-one forgets their first total eclipse; and many are led to seek another, and another.

Tourist expeditions to total eclipses of the Sun are not new, though air travel has made them easier. In 1898 several thousand inhabitants of San Francisco and Sacramento took excursions by train to the site of totality, and in 1900 more than 20,000 left Madrid by the same means to reach the line of a total eclipse that crossed the Spanish peninsula. It was, however, the splendid event of June 1973 that initiated the modern trends in eclipse tours. No fewer than seven cruise liners sought clear skies off the bulge of Africa, and overland tours converged on Kenya to combine game parks with views of the celestial phenomenon.

Francis Baily and his 'beads'

Born in 1774, Francis Baily was a man who read extensively the literature on solar eclipses. He made his career in the Stock Exchange, worked tirelessly, and became wealthy enough to retire at the early age of fifty-one. Yet long before his retirement, Baily devoted much of his spare time to scientific matters, and to astronomy in particular. He was one of the founders of the Royal Astronomical Society, in 1820, and as early as 1811 published a thoughtful analysis of the ancient classical references to solar eclipses. From this work grew his ambition to see at first hand a total eclipse. He did so, but not until he was sixty-eight years of age, and only two years before his death. Prior to that he was able to observe the less spectacular varieties of solar eclipses.

In 1820 business commitments prevented Baily travelling to the path of an annular eclipse that crossed Europe. The poet Wordsworth witnessed that eclipse from Lake Lugarno in Switzerland, and penned a rather dreary poem about it, which alone might have deterred a reader from making the effort to see one. Baily saw the event as a partial eclipse from his home outside London. He then waited patiently for sixteen years to pass.

The annular eclipse of 1836 was visible from Britain. Baily journeyed north to the outskirts of Jedburgh, taking with him his $2\frac{5}{8}$-inch (6.8-cm) telescope, and set it up in the garden of one of his friends. Clear skies favoured his site, and Baily watched the entire event. As the Moon glided fully within the Sun's disk, he was struck by an unexpected sight. The slight irregularities of the Moon's limb – the walls of craters and mountain ranges seen in profile – stretched out to link the edges of the Sun and Moon, and so to delay the completion of the narrow annulus. Between the black peaks, intense spots of sunlight shone through. After a few

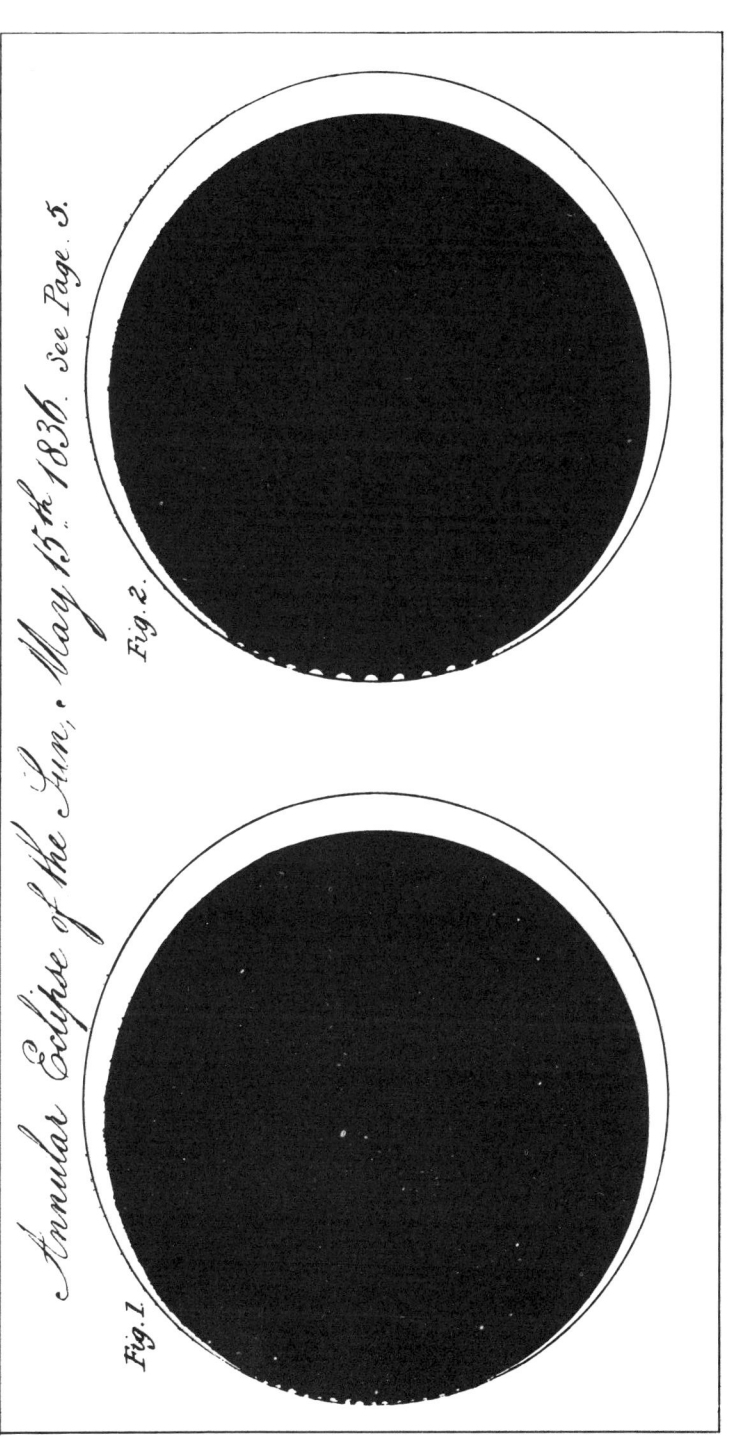

Figure 1. Baily's original drawings of the annular eclipse of 1836, showing the string of beads at the left edge, were reproduced in the Memoirs of the Royal Astronomical Society.

seconds the black links were severed, and the eclipse became truly annular.

Baily wrote a scholarly account of his observations for the *Memoirs of the Royal Astronomical Society*. He searched the literature and found a number of earlier references to the same phenomenon, at both annular and total eclipses.

The earliest-known reference to Baily's beads, and in particular to the so-called diamond-ring effect when one bead is extremely bright, dates from 1133 and the chronicles of the Frenchman Honorius. Although Baily's description was not the first, it remains the best, and to this day the appearance of a set of bright spots of sunlight, strung like beads on a necklace several thousand kilometres long, is called Baily's beads.

Six years later Baily saw his total eclipse. He viewed it, at his request, locked alone in a room high in the University of Padua, in Italy. On that occasion Francis Baily learnt what an emotional experience a total eclipse is, and how much more wonderful than the annular variety.

George Airy

George Airy was Astronomer Royal in 1842. A petulant man, myopic in both senses of the word, loathed by most of his staff, he none the less sought projects to involve himself in which would bring new directions to Greenwich. Airy was impressed by Baily's description of the beads and decided to take himself off to the 1842 eclipse for a better look.

Airy was rewarded with a fairly clear view of totality, and he too found the experience very moving. He had not, however, done his homework. When the corona and prominences became visible, Airy was taken by surprise. Just what he had expected we know not, but certainly not what he witnessed.

The scientific literature in that year already contained a great many accounts of solar eclipses. Two in particular were detailed and informative. Halley had made a very careful study of the 1715 eclipse – far better than Airy's account more than a century and a quarter later – and the Swedish scientist Celsius had compiled a marvellous description from meticulous observations of the 1733 total eclipse made by local clergymen.

Perhaps it was to Britain's advantage that their Astronomer Royal had ignored the literature. Airy was so impressed by what he saw that he mobilized Britain's astronomers to study the issues raised. In so doing he initiated a golden age of exploration and discovery, an age in which British astronomers were to claim most of the line honours.

'Cosmically it is of the utmost importance,' wrote Airy, 'to determine whether these appearances are attached to the Sun or to the Moon'. These words seem strange to us now, comfortable in our knowledge that both corona and prominences are manifestations of the outer, tenuous layers of our star. In the middle of the last century these simple facts had yet to be established. A sizeable minority believed the rose-red prominences to be jets of gas escaping from the Moon's surface, even though a careful reading of Celsius' collation clearly indicated the contrary. Several of the Swedes had seen the prominences steadily disappearing behind the Moon's limb as it glided across the Sun, whilst other prominences were revealed at the opposite edge. The corona aroused greater discussion, with comparable numbers of astronomers arguing for a lunar or a solar origin, and a few believing it to be partially or entirely caused in the Earth's atmosphere. Kepler, in 1620, was the first to suggest that it was the Moon's atmosphere we see.

Although total eclipses of the Sun occur, on average, once every seventeen months, they show complete disdain for the abodes of astronomers, preferring to cast their shadows across the vast oceans or into remote corners of the globe. In the middle of last century, travel to remote spots was prohibited by both practical and financial restrictions; the British had to have patience. Their next opportunity came in 1851, in Scandinavia. Making allowance for the risk of cloud, Airy mounted a major expedition and despatched observers to a score of locations. Some had it clear, and from their accounts, backed up by the first use of photography, the solar association of the prominences was finally deemed correct. With that demonstration came the recognition that these giant flame-like formations utterly dwarf our planet.

A further nine years had to pass before British astronomers again sought the Moon's shadow, in 1860. In the interim an enterprising team of Chilean astronomers made some observations of totality from Peru. Soon other nations began to be interested.

Expansion – Adventure – Hazards

Beginning with the 1860 eclipse we see a dramatic surge in

interest. Suddenly total eclipse observations became fashionable astronomy, making a welcome change from measuring the positions of myriads of stars, still the major work in the science. It became a matter of national pride to sponsor eclipse research, and with the ever-improving facilities for travel, expeditions began to sally forth to odd corners of the globe. The dates of eclipse expeditions began to cluster more tightly together: 1868, 1869, 1870, 1871, 1872, 1874 .....

These expeditions were not without their adventure. Often the astronomers had to travel to essentially uncharted lands, or found on arrival that the maps they had were quite wrong. Some expeditions effected almost as great a study of the lands they visited as of the Sun. The 1883 trip to the Caroline Islands was one such example. One American expedition, in 1860, journeyed by stagecoach, paddle steamer, and birchbark canoe into the little-explored waterways of Saskatchewan. A zoologist named Scudder, who later became a famous entomologist, classified the only wildlife he could find; several distinct species of mosquito. Scudder later lamented that the expedition suffered many hardships only 'to sit in a marsh and view the eclipse through clouds'.

The 1870 event, in North Africa, was particularly memorable. At the time two great solar astronomers were vying for honours in the field: Pierre Janssen, of France, and the Englishman Sir Norman Lockyer. Lockyer was shipwrecked on the Sicilian coast, but escaped not only with his life but also with his scientific equipment. Janssen found himself stranded in Paris which, at the height of the Franco-Prussian war, was a city under siege. Undaunted, he escaped by hot air balloon only to be greeted by a solid cloud cover at his chosen site. Janssen it was who, at the 1868 eclipse in India, had demonstrated that the crimson of the prominences is due to the Sun's most abundant gas, hydrogen. Lockyer bettered this by adding helium to the list, an element which, though predicted, had not been found on Earth at the time.

The corona became the major field of study. If it were of terrestrial origin it must appear different from different locations, hence the need to spread out the observing parties. But even when a solar origin was sure there remained questions of how quickly it changed. From one eclipse to the next it changed completely, though it began to show a rhythmic pattern corresponding to the eleven-year sunspot cycle. Did the corona vary over a period of a few hours? To answer that question it was necessary to despatch observers to opposite ends of the eclipse track, increasing the difficulty of a successful expedition. A comparison of the corona seen from the two sites demanded excellent photographs of it. Not until 1900 was photography clearly proved superior to the trained artist; when we compare sketches of the corona made by observers standing almost side by side we appreciate how impossible it was to answer that question before the turn of the century.

Photography in turn meant larger and more sophisticated equipment. Telescopes of ungainly size and weight had to be transported to the ends of the earth. On arrival, the astronomers had not only to construct mountings and shelters for these telescopes, but also had to effect photographic darkrooms. They faced enormous difficulties. The temperatures were often higher than the chemicals could tolerate; the humidity prevented drying of the emulsions; and hordes of insects had to be kept off the photographic plates because they stuck and so damaged the irreproducible images.

The expeditions escalated in size and complexity, and so in cost. Manpower alone was enormous – an estimated three man-years per expedition for the astronomers alone. If the travel had to be to far-flung places – Japan, India, or the Pacific Islands, for instance – the observers might easily be away for three months. No sooner would they return from one trip than they began planning the next. For some it became almost an annual ritual, often gruelling in the extreme. 'Benkoelen has a warm reputation', wrote one after the 1926 eclipse in Sumatra. 'The damp heat day and night reduced even the strongest to perspiration and profanity.' There was the ever present risk of disease, brought home poignantly when the leader of one British expedition, to the West Indies in 1889, died of dysentery. On another occasion the astronomer was taken so ill that his wife took on the enormous responsibility of processing his photographs. Some expeditions were provided by the local authorities with doctors carrying snake-bite sera.

'like true love . . . never runs smooth'

The expeditons may have been hard, but they were also fun. We can read the enjoyment in the words of those few who had a gift with the pen. One such person was H. H. Turner, an Oxford don who wrote extensively for the journal *The Observatory*. Here are a few of the comments Turner sent back.

1896 Japan: 'There is a fine old Japanese gentleman living on the hill, who was once a Samurai and never drew his sword without killing his man; now he manufactures iodoform from seaweed.... He and his married daughter have the most beautiful manners, and they send us gooseberries and herrings, and the kindest messages – as, for instance, that they were sorry to think they could do so little for us but perhaps it might be useful to know that they had a servant who could starch linen.' But also: 'My poor golf clubs have not been taken out of their case; I never saw a more abjectly hopeless country for the game.'

1898 India: 'There was an elephant put at our disposal by the local Rajah, but we did not care much for elephant-riding after the first day: it is too jolty'.

1905 Egypt: 'The Survey Office is one of the more modern buildings in Egypt, having been built at Giza in 1900 (A.D.) There are in the same neighbourhood some older buildings, pyramidal in shape, which have sometimes attracted attention. We went there by electric car and donkey, the first method of progression being preferable'. On the other hand: 'the water in a disused bath, in which a few thousand bricks have been soaked for building our piers, is preferred for drinking to the unadulterated Nile itself.'

Another astronomer who could delight the reader with his descriptions was the Australian F. K. McClean. Of the desolate Flint Island in 1910 he wrote: 'Several members soon became skilled in the art of boot-making with pieces of carpet owing to the sharpness of the coral, considerably assisted by the ravenous propensities of the various crabs. Our steward had part of his trousers eaten up in this way'. Apparently, tethering these giant crabs was to no avail. If anchored by rope they would quickly bite through it; if wire was used the crabs calmly bit off their tethered leg.

Not all expeditions were successful, of course. Though they would pick the sites offering the best climatic chances, they could not always be right. 'The most exasperating part of it all', wrote the great American astronomer E. E. Barnard, 'was, perhaps, the fact that if instead of going away off into the interior to search for good weather, we had simply gone ashore from the gunboat that carried us to Sumatra, and put up our instruments on the sea-shore near it, we should have had successful observations.' Less predictable problems included the eruption of a Japanese volcano whose smoke blotted out the Sun in an otherwise clear sky; and two

Figure 2. This scene is typical of eclipse expeditions around the turn of the century.

policemen, sent to control onlookers, who stepped in front of a telescope at the moment of totality. E. W. Maunder opened his book on the 1900 eclipse expedition with these marvellous words: 'The course of Total Eclipse Expeditions, like that of true love, seems never to run smooth. Of the three which the British Astronomical Association has organized, the first was baffled by cloud, the second was hampered but not thwarted by plague, and the third was hindered but not beaten by war.' Turner summed it up in a very quotable statement: 'Disappointment is so often the lot of the astronomer, that if he cannot cheerfully contemplate reverses it is not for want of practice.'

The scientific goals evolved rapidly, which is why eclipse expeditions continued for so long and indeed still continue to this day. There were many subtle measurements to be made, and very little time in which to make them. Eclipses lasting more than three minutes were rare. A cool head and a well-rehearsed experiment was required, and legion are the tales of astronomers so awed by the spectacle that they failed in their tasks. Expeditions became almost military in their approach. In 1911, for instance, an Australian team in Tonga used the following count-down:

−10 mins:	bugle 'Rouse up'
−5 mins:	bugle 'Alert'
−42 secs:	3 whistles
−9 secs:	2 whistles
−4 secs:	1 whistle
zero:	voice 'Go'

Total Eclipse and Einstein's Theory of Relativity

Arthur Stanley Eddington has been described as the greatest astrophysicist of his time. In the 1920s he became the first person to understand the internal constitution of stars, and to fathom out how they produce such prodigious energy. A decade earlier, in 1912, he began his astronomical career on the eclipse circuit, only to find the work suspended by the First World War.

During the war Eddington evaded conscription. He would in fact, have elected to serve his time in prison as a conscientious objector, but was spared that ignominy largely by the efforts of Sir Frank Dyson, the Astronomer Royal. During the war years Eddington studied Einstein's newly published theory of relativity until he understood its every detail as well as Einstein himself. Eddington, who believed implicitly in the truth of the theory, began to explore the new view of physics it allowed, and this helped him considerably in understanding the stars.

Most physicists and astronomers lacked a deep grasp of relativity; most sought proof of the theory. Eddington knew that one simple test could be made during a total eclipse of the Sun. Relativity predicted that light, when passing near an object as heavy as the Sun, is deflected very slightly from the straight line path we are all taught that it follows. Stars near the Sun appear to be pushed away from the Sun's centre, much as a stick appears to bend when dipped in water. The amount is tiny – only 1.75 seconds of arc at the very limb of the Sun, and progressively less further away. Stars near the Sun can been seen only during a total eclipse, of course.

A measurement of the deflection is most readily made when the Sun sits in a crowded field of stars. In late May each year it passes through the star cluster known as the Hyades, which surrounds the bright red star Aldebaran in the constellation Taurus. Around that date is the best time to attempt such a measurement. Remarkably, the 1919 eclipse was due to fall on May 29. Dyson persuaded the authorities to allow Eddington to plan an expedition to measure the deflection of the stars. Eddington felt the whole business to be unnecessary, since nobody could possibly want to confirm a theory as beautiful and as right as Einstein's. None the less, he went along with Dyson, was spared the draft, and led the 1919 expedition.

The expedition lasted four months, because it was essential to photograph the Hyades at night with exactly the same equipment in the same place in order to compare the precise positions of the stars with and without the Sun in their midst. Another couple of months passed while the precious plates were measured and re-measured, and the results digested. Finally, in November 1919, the conclusions were released at a joint meeting of the Royal Society and the Royal Astronomical Society. Newspapers across the country and around the world headlined the news: Einstein was right; relativity was right; British astronomers had proved it so.

Too routine

Eclipse expeditions continued. Better measurements were made of the deflection of starlight, without changing the outcome. Other subtleties of the Sun were explored and measured, and as recently as 1983 an expedition used a total eclipse as a means of measuring the diameter of the Sun with great precision, using a technique thought up by Edmond Halley. Observers were spaced along the edges of the shadow band, and were able to delimit the shadow according to which of them did or did not see the Sun completely covered.

Each successful expedition has been another triumph. Yet, somehow, the business has become altogether too routine. The truly romantic age of the shadow chasers ended in 1919.

GARRY E. HUNT

Future Exploration of Mars

1. Past Missions

Mars has always been the one planet of our Solar System that has been thought to be potentially habitable. Venus, although Earth-like in size is quite inhospitable. The surface environment comprises a crushing pressure of a hundred times that of the Earth and an oven-like temperature *everywhere* of 737 K, hidden underneath ubiquitous sulphuric acid clouds, which have a yellowish-lemon hue to the external observer. Clearly, there is little likelihood that any human can explore the surface of Venus, which is more like a living hell. Only Titan, the satellite of Saturn, resident in the cold depths of the outer Solar System, which currently resembles the Earth in deep freeze, could perhaps one day possess some form of life. But, it is unlikely to have more than a complex chemistry at this time. Consequently, Mars is most likely the one planet we can hope to understand as well as our own and perhaps even live on in the near future.

There have been a number of space missions to Mars (*see* Table 1), which have progressively increased our knowledge of the planet. The early missions suggested that Mars was a dead Moon-like object. We now know it to be a geologically, meteorologically and just conceivably a biologically active planet.

The Viking mission with its two orbiters and landers has provided the most important discoveries. However, we should not measure the success or failure of this mission simply on the biological investigations, which still remain unresolved, but on the contribution it has made to our total understanding of planetary processes. There is no doubt that Viking has been a resounding success with fundamental information obtained on the cause of the Martian channels, the nature of the polar caps, the size of the

TABLE 1
Past Missions To Mars
(The year refers to the time of the spacecraft encounter with Mars)

MARS 1	USSR	1963 Flyby of Mars at a distance of 193000 km on June 19, 1963, but contact lost on February 21.
MARINER 4	USA	1965 First successful flyby of Mars on July 14, 1965 at distance of 9850 km, producing 21 photographs.
MARINER 6	USA	1969 Successful flyby on July 31.
MARINER 7	USA	Successful flyby on August 5.
MARS 2	USSR	1971 Entered orbit around Mars on November 27. First capsule to reach the surface crashed; no data.
MARS 3	USSR	Entered orbit around Mars on December 2. Capsule landed but transmitted for only 20 seconds.
MARINER 9	USA	First successful Mars orbiter; returned 6876 pictures.
MARS 4	USSR	1974 Mars flyby on February 10, 1974.
MARS 5	USSR	Entered Mars orbit on February 12, 1974.
MARS 6	USSR	Capsule landed on Mars but contact lost prior to touchdown.
VIKING 1	USA	1976 First successful Mars landing on July 20; separate orbiter spacecraft.
VIKING 2	USA	Second successful Mars landing on September 3; further orbiter.

ancient atmosphere, which collectively have helped to provide a more complete understanding of the complex issue of the past, present and future characteristics of the Martian atmosphere.

2. Challenging Problems For the Future

Mars, the Red Planet, is intermediate in size between the Earth and our Moon. Part of the planet's surface resembles the Moon and shows massive impact basins, cratered highland regions and extensive flooding by lavas. Even more exciting are the regions which resemble the Earth; the mountains, volcanoes, dried-up river beds, desert sand dunes, seasonal polar caps, cloud systems, and other weather patterns and evidence of climatic change. Apparently, Mars has evolved to an advanced stage, well beyond the development of the Earth. It is possible that the Martian internal heat engine powered by the decay of radioactive elements, is still active. This would create active volcanoes and the exhaling of gases into the atmosphere. This would be a sensational discovery to make.

There is evidence of a measurable Martian atmosphere composed of carbon dioxide, with small amounts of nitrogen, argon, water vapour, and some other inert gases, and a planetary surface whose temperature may locally rise above the freezing point of water from the current ice-age temperatures. This inevitably raises the question of whether Mars could have developed indigenous life. Indeed the controversy over this topic has simmered and boiled for centuries. There is, alas, no conclusive answer yet.

The Viking landers sought to resolve this major problem which is at the very foundation of our knowledge and enquiry. From their positions on the Martian surface, thousands of kilometres apart, a number of experiments were carried out with samples of locally collected Martian soil treated with nutrients. Unfortunately, these biological experiments proved neither positive nor negative, so currently, there is no evidence of any Martian life. Even more discouraging is the clear result from the organic analyses of the soil samples which suggest that there is essentially no organic material in the soil above the parts per billion threshold.

This result was a major surprise. Even in the absence of life, the impact of carbonaceous meteorites on Mars over several aeons should have supplied a substantial amount of non-biological organic molecules to the surface material. Perhaps this material has been destroyed through the passage of time by the intense ultraviolet radiation, since, unlike the Earth, the Martian surface layers are not protected by an ozone layer. But there are many problems with the analyses carried out on Viking. Were they the

Figure 1. The bare, rocky surface of Mars as seen by the Viking landers. NASA.

correct experiments? Why were we looking for simply a carbon-based life form? Surely it is possible that life could have existed in the geological past so that any evidence would be difficult to find from samples of the top soil which has been constantly mixed and moved around by huge dust storms? Clearly, we need a new approach to this major problem and the ability to roam around the planet with robots or even astronauts to carry out the experiments in a wide variety of locations. Certainly, two landing sites may not be totally representative of the entire past and the manner in which it has developed.

Irrespective of whether Mars has ever been an oasis for life, the nature of this planet and the distinct possibility that colonies could be set up there, makes it a compelling target for in-depth exploration and exploitation in our international space programmes.

A wide range of geological processes has operated on the planet to produce landscapes which are both familiar and alien to our understanding. Individual shield volcanoes that would stretch from Boston to Washington, DC, with similar properties of the Hawaiian volcanic chain; a canyon that would extend from New York to Los Angeles, with seas of sand dunes that girdle the entire north polar region.

Mars is also a perfect laboratory for studying planetary weather and climate systems. The diurnal and seasonal cycles on Mars are remarkably similar to those on Earth. However, the very thin Martian atmosphere characterized by a surface pressure of only 7 mb, compared with about 1000 mb for the Earth, means that the Martian atmosphere responds rapidly to the heating by the Sun and the heating and cooling of the surface layers. The Martian condensation cycle involves carbon dioxide which provides a seasonal covering to the polar caps in addition to the more familiar processes involving water vapour. These are fascinating contrasts to the terrestrial systems. However, even with the current set of observations of Mars from the past missions which were not optimized for meteorology, there is every indication that the exciting and complicated Martian weather systems are capable of being understood and providing useful information for the benefit of improving our understanding of terrestrial phenomena.

There are also some paradoxical observations of Mars which provide clear evidence of past climates on the planet, which may have been triggered by mechanisms that have affected our own planet. On the Earth, we have the evidence of climate change from the ice ages, changing shorelines, and the extinction of particular species. On Mars, there is evidence through regional flooding, glaciation, and periodic polar sedimentary layering. The possible common mechanisms which have contributed to these planetary changes are variations in the solar energy, orbital variations of the planets, episodes of volcanic activity, and asteroidal impacts. The detailed comparative observations from the Earth **and** Mars could make progress toward understanding an issue of both intellectual and practical importance.

We must remember, that Mars is not alone. Indeed the planet has two strange but tiny moons, Phobos and Deimos, which were observed at close range during the Viking Mission. They are both very irregular bodies; Phobos measures $27 \times 21 \times 19$ km and Deimos a mere $15 \times 12 \times 11$ km. Their surfaces are covered in craters with a range of sizes which suggest they may both be the surviving fragments of larger bodies. Surprisingly, Phobos is dominated by a large 8-km crater, Stickney. But the origin of the moons is unknown. Were they formed from the same material as Mars or are they captured? Furthermore, in contrast to Phobos, tidal forces are pushing Deimos away from Mars. This is a further important question to address in any future mission to the Martian system.

3. Future Missions

There are several Martian missions being planned by both the Soviet Union and the United States as essential parts of their planetary programmes. The ultimate goal of these separate programmes is to explore fully the planet and its moons; to answer, unambiguously, the question of past or even present life on Mars; and then finally set up colonies on the Red Planet. This may involve sophisticated robots on the surface of Mars, soil sample return missions, in addition to usual orbiting spacecraft equipped with the latest instruments, communication and control systems. Consequently, to achieve these objectives will extend the current capabilities of current technology, which in itself is a mechanism of using all the basic skills of our rapidly evolving civilization.

The Soviet Phobos Mission

The Russian space programme will contain the next mission to the Martian system. In July 1988, they plan to launch two probes to Mars and the moon Phobos. The mission is in many ways a follow on to the very successful VEGA space mission which sent probes to both Venus and Comet Halley. Once more it is a very international mission, which is a very pleasing development of Soviet space missions in recent years. There are hardware contributions from Austria, ESA, Finland, France, Sweden, Switzerland, West Germany, as well as the other INTERKOSMOS members, namely, Bulgaria, Czechoslovakia, East Germany and Hungary. American and UK scientists may also be involved in the mission as the schedule develops.

After the 200-day cruise to the planet, the three axis stabilized spacecraft will be placed into a 4200×79000-km, three-day period elliptical orbit. After 25 days the periapsis will be raised to alter the orbit to a 9700×79000-km path for 30 days. The Phobos encounter is achieved by first moving the spacecraft into a circular orbit of radius 9700 km and with a period 8 hours before moving into a 9400-km circular orbit of period 7.6 hours which will take the spacecraft to within 30 km of Phobos. The main spacecraft is then manœuvered to a distance of 50 m of the tiny satellite and two surface landers are finally deployed to remain active on the satellite surface for about a year. It is intended that this deployment phase of the mission will be fully televised. This must make compelling viewing for everyone.

A wide range of experiments will be conducted at Phobos, on the surface of the satellite and at Mars. The Mars–Phobos mission represents a quantum leap forward in terms of Mars exploration, just as the Viking mission did more than a decade ago. The results from the thirty-one experiments will surely rapidly advance our knowledge of the Martian system and also prove the technological capabilities in preparation for robot and possibly manned missions of the future.

NASA Mars Observer Mission

In comparison with the Phobos mission the next NASA mission to Mars may not seem quite so glamorous. However, it does have a very important role to play in advancing our knowledge toward

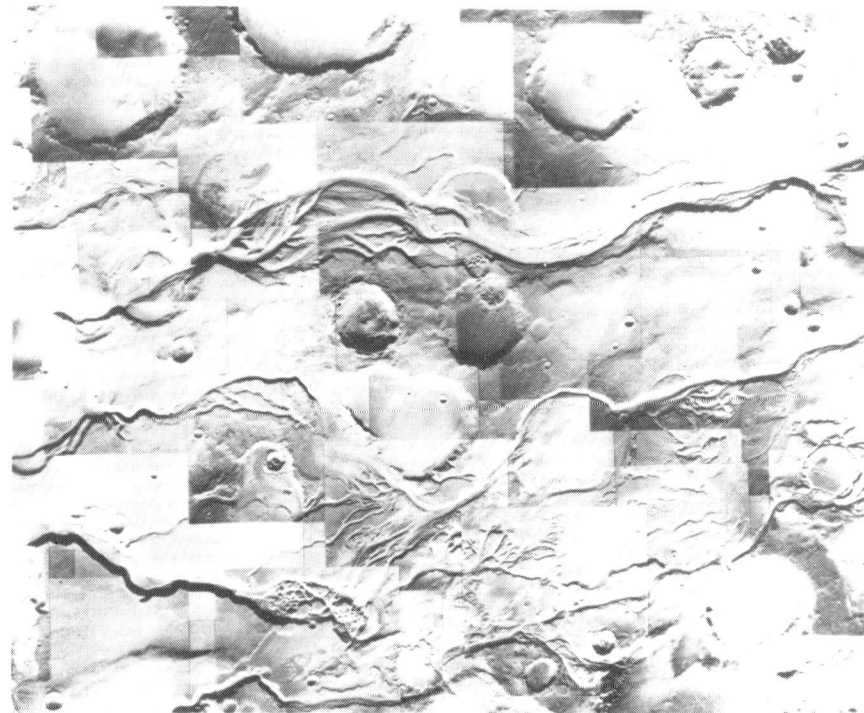

Figure 2. This is a mosaic of the lower portion of a complex of channels on Mars called Mangala Valles. The channels appear to have been cut when large amounts of water were released from beneath the surface. Visible are many streamlined islands and interior channels, typical of erosion by rivers and streams. Also visible are many examples of sediments originally laid by ponding and then re-cut by later flows. An example of chaotic terrain can be seen at upper right. Scientists believe it is a result of withdrawal of ground water or ground ice during a melting period. Some impact craters in the scene have been highly modified by erosion and wind-deposition of the surface material. The changes appear to be related to the process that formed the valleys. Other craters in the area are pristine and, therefore, relatively younger. Smaller valleys, some formed by the movement of materials within pre-existing, larger channels, can be seen cutting through the scarp or cliff at extreme right. There may be large amounts of water-ice still held in the surface materials in areas like this where flooding has occurred.
Reproduced courtesy of, and copyright by California Institute of Technology, NASA.

possible soil-sample return missions at the end of the century. The current US shuttle launch problems now mean that NASA cannot launch this spacecraft until 1992. However, when the spacecraft is in orbit around the planet, it will provide continuous observations of the atmosphere and surface properties to address the basic questions remaining from the Viking mission; namely determining the global mineralogical characteristics of the surface, the global topography and magnetic field, the characteristics of the magnetic field, and the meteorological systems. In this way the US probe measurements are complementing the Soviet mission.

THE NEXT STEP

The fundamental questions about Mars do require detailed analysis of the soil which will not be carried out by either of the currently planned missions. Certainly such missions are being discussed even without any available funds. Consequently, we can imagine that the next step beyond the Phobos and Mars Observer Mission, must be return to Earth with samples of soil for analysis. It is likely, however, that the analysis would be carried out in Earth orbit so that there would then be no danger of contamination of either the sample or our own frail planet. Spacelab would be the ideal place for these analyses.

Beyond this mission we must expect soon that a manned mission to Mars will take place. The Soviets have frequently stated this to be their ultimate intention. In the US the authorities have recently set up a Committee involving several of the Apollo astronauts to spearhead the US toward the next quantum leap in space exploration. Already the Soviet cosmonauts have satisfied the endurance aspects of the flight time. Such a mission would involve about 210 days for the outward trip, a three-week landing period and a similar period for the return leg. Manned missions to Mars are well within our technological capabilities and are a necessary part of our evolving space capability. We could well see such a mission in the next 10-20 years.

R. C. MADDISON

Background to the Big Bang Theory of the origin of the Universe

The use of a title like 'The Origin of the Universe' would seem to be rather arrogant, since it is obviously not possible for us to have access to any authoritative account of what actually happened at that very first instant of time. However, it is possible to select from the vast number of speculations on the subject those that seem to be supported by observation. Many of the earlier ideas are based on myth, superstition, and vested interest and have been put forward by equally unqualified persons in most dogmatic fashions as if they had been revealed in some privileged way by a supernatural power.

We are all familiar with the teachings of groups who 'know' the truth – but who are, strangely, unable to present any direct evidence of their sources of information. I would like to review some of the scientific evidence that throws some light on to this whole area of dark speculation. This scientific approach to the problem differs from other approaches in that evidence is measured and can be checked. Anyone is free to challenge and interpret the evidence; to put forward theories which can be used to predict circumstances which may again be checked by observation and experiment, and in this way a body of observed 'fact' may be established. Scientific Law, unlike Civil Law, is not to be broken, for if it is broken, it ceases to be law. Civil Law is unnecessary if it is unbroken but scientific law stands until it is successfully challenged, whence it must be modified so that it encompasses the new evidence. Even though we may still not be able to probe the distant past by experiment to find what happened, we may be able to suggest with greater certainty what did not happen and this must surely lead us closer to the truth.

First we must define more clearly the nature and scale of our problem.

Behaviour of the Universe

The idea of 'creation' is fundamentally alien to science. We have learned that, according to measurement and observation of a very high order of accuracy, all things obey laws of conservation. Nothing can be destroyed or created. It may change from one form into another, but it cannot be totally eliminated or conjured up from empty space. The creation or destruction of the Universe are situations that violate this cherished principle of conservation, and it is clear that presently available scientific method and techniques cannot deal with such events. We have no evidence that the laws of science have not changed or perhaps evolved over aeons of time. We can, however, study the behaviour of the Universe as we see it, and probe into the past under the assumption that the laws of science have not changed significantly over the period we can observe. The astonishingly consistent picture that emerges from this treatment tends to justify this assumption, and it does at least enable us to discern overall trends in the behaviour of the Universe and put forward some theories to account for them.

What then do we mean by the term 'Universe'? A loose definition might be that it is 'everything that exists; all that we can see around us; everything that we can sense with our most sensitive detection techniques'. It incorporates the sub-microscopic structures that we find on the atomic scale of dimensions out of which matter itself seems to be made, and also the gigantic systems of stars on the scale of galaxies whose sheer size defeats the imagination.

These things are all subject to scientific investigation, and it appears that the system of scientific law that has been built up over the centuries applies equally well to the very small and the very large.

Structure of the Universe

The very earliest ideas on the structure of the Universe were based on common sense. The Earth was seen to be flat and of indeterminate extent, although it was thought to be finite. No-one had successfully returned from an extended journey on the big seas – presumably they had sailed over the edge and fallen into oblivion. No movement of the Earth could be felt; so, obviously, it must be fixed. Thus the flat earth was regarded as central in the scheme of things, and the various lights that move around the sky in a complicated way were of secondary importance and put there merely for our convenience. The notion of a designer and builder arose without difficulty.

It was the cyclical behaviour of the Sun, Moon, stars and planets that eventually led the Greek philosophers of the fifth and fourth centuries B.C. to construct a picture of the Universe in which the Earth was spherical and moved around the central Sun in an annual orbit. The view that the Earth is spherical was accepted widely at the time, but the prevailing opinion that it was the centre of the Universe was to remain dominant until the late sixteenth century.

Aristarchus (310-230 B.C.) actually attempted to gauge the relative distances of the Sun and Moon, and although the method he used was faultless, his equipment was crude and incapable of the necessary precision. Nevertheless, he was able to demonstrate that the Sun is much more distant than the Moon although his actual estimates were grossly in error.

The first major advance in setting the scale of things was made about 195 B.C. by Eratosthenes, who was Librarian in the city of Alexandria. He knew that in the town of Syene, near the modern Aswan – some 500 miles to the south of Alexandria – the noon Sun on the day of the summer solstice was directly overhead, so that the solar rays shone directly into the bottom of a well. But in Alexandria, at this time, the shadow of a large upright pillar showed that the Sun was $7\frac{1}{2}$ degrees from the zenith or overhead point.

If, therefore, the Sun is very distant, then its rays would be striking Alexandria and Syene from the same direction and the $7\frac{1}{2}°$ angle must be caused by the curvature of the Earth. Five hundred miles across the Earth's surface leads to $7\frac{1}{2}°$ difference in the apparent direction of the vertical. How many miles, then, are necessary for a 360° deviation? Approximately twenty-four thousand, so this must be the circumference of the Earth. There is still some discussion about the actual units of distance used in those days, but it is clear that this figure is remarkably close to the modern accepted value.

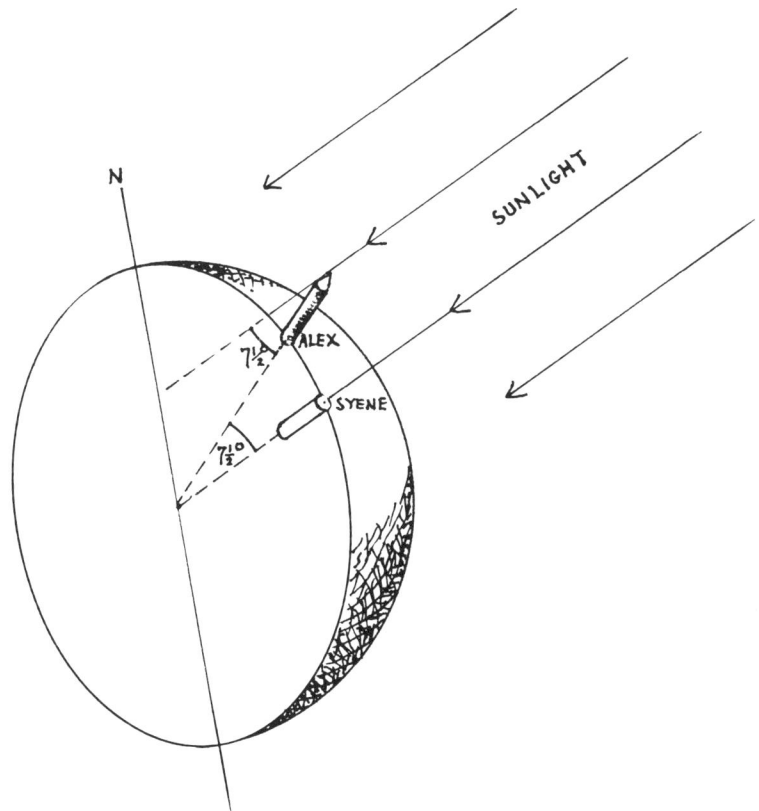

Figure 1. Method used by Eratosthenes to estimate the size of the Earth.

The idea of a spherical Earth was, of course, not new, but the measurement of its size was. At that time the Earth was still regarded as the most important object in the Universe and therefore its position was central in its structure. This erroneous geocentric theory was very well established, and received further weighty support from Claudius Ptolemy in A.D. 150 when he published a series of volumes called the 'Almagest' which was regarded as the definitive text of the period. In it the planetary motions were represented by a complicated system involving large spherical shells, centred on the Earth, called 'deferents', along which moved the centres of smaller shells called 'epicycles' which carried the planets. By carefully choosing the various periods of rotation, this 'Ptolemaic' system could be made to work satisfactorily within the accuracy of the measurements of planetary positions available at the time.

The theory was to remain more or less unquestioned until 1530, when Nicolaus Copernicus (1473-1543) recognized that a more accurate description could be made by having a Sun-centred system in which the planetary orbits were simple circles which were not themselves concentric. The Earth thus became one of the Sun's family of planets – a degradation in importance that was received with dismay in many quarters. Galileo Galilei (1564-1642) re-enforced this concept with the first telescopic observations of the sky, culminating in the discovery of phase changes of the planet Venus, which can only be explained if the planet actually orbits the Sun as seen from Earth. Shortly after Galileo's discovery Johannes Kepler (1571-1630) published the results of some twenty years of his own study of planetary motions based on observations obtained by Tycho Brahe at his observatory on the Danish island of Hven. He had accidentally stumbled upon three laws which described the motions of the planets with very great accuracy. The first was that each planet moves in an elliptical, rather than a circular, orbit around the Sun; the Sun lies at one of the foci of the ellipse. The second law describes the way in which the speed of a planet varies with its position in orbit, and imagines a line joining the planet to the Sun sweeping out equal areas of space in equal time intervals. The third law is an arithmetical relationship between the average length of this imaginary line (R) and the amount of time (T) required for the planet to complete one orbit. It appears that the ratio R^3/T^2 has the same value for each planet. These three laws of planetary motion revealed for the first time the nature of the force holding the Solar System together. Like all other forms of energy, the strength of gravity diminishes as the distance of the source increases in accordance with an inverse square law. Thus doubling the distance reduces the effect to one quarter of its former value.

In the hands of Isaac Newton (1642-1727) these laws were to be developed and incorporated into a comprehensive theory of gravitation that could explain orbital motion in terms of a Universal Force of Gravity – an idea that was to transform our view of the Universe.

Impact of Telescope on theory

By the beginning of the eighteenth century the heliocentric theory was widely accepted. The Sun had replaced the Earth as the centre of importance in the scheme of things, and the stars were seen to form an obviously very distant background of totally unknown extent.

The only clearly seen structure in this starry background is the Milky Way. Any small telescope reveals that this ribbon of misty light, extending across the sky from horizon to horizon, is composed of myriads of faint stars. Travellers around the Earth see it as a continuous belt around the entire sky, and its presence surely signifies something about the overall shape of the Galaxy.

Thomas Wright, in 1750, was the first to suggest that perhaps most of these stars are faint because of their great distance. Perhaps there is greater depth in the Universe in the direction of the Milky Way than in any other direction?

Since the Milky Way seems to be almost equally dense in all directions, this seems to imply that the system of stars is roughly disk-shaped, like a giant grindstone wheel with the Solar System somewhere near its centre.

Making the same broad assumptions as Wright, but taking account of the differing brightnesses of the stars, William Herschel (1790) carried out a painstaking and thorough survey of how the stars are distributed across the entire sky. His conclusion was much the same as Wright's, namely that all the stars make up a flattened, disk-like system having the Sun near its centre. Neither of these observers realized the extent and importance of the clouds of interstellar material that limit our view in some directions. They believed that they were seeing right through to the extreme edge of the entire Universe.

By the middle of the nineteenth century telescope design had progressed to the point where giant instruments, such as the Earl of Rosse's 72-inch aperture reflector, were being aimed at the stars.

Several peculiar whirlpool-like objects were studied which were initially thought to be great rotating clouds of gas, huge vortices that were part of the foreground and probably not very far away.

The introduction of photography allowed observations of much improved sensitivity, because light could be stored during long exposures, and the result was the discovery of many more of these 'nebulæ' all over the sky. It seemed strange, though, that there appeared to be a 'zone of avoidance' matching closely the Milky Way in which none of these objects could be found. We know now that this effect is again the result of the large quantities of interstellar material concentrated in the plane of the Galaxy that limits our view as effectively as a bank of fog.

Detailed study of these spiral nebulæ showed that they can be seen displayed at all different angles; some are even seen edge-on, and it was not long before the suggestion was made that maybe they are not simply local spinning gas clouds – perhaps they might be island systems of stars completely separated from the Milky Way. Unfortunately no good method of estimating distances was available at the time, and the true nature of these objects was not resolved until the beginning of the twentieth century.

The period – luminosity law

The key to the problem was found in 1912, when Henrietta Leavitt, working at Harvard, noticed a relationship between the luminosities and the periods of a group of stars known as the Cepheid variables.

Luminosity is a measure of total power output, or the intrinsic brightness of a star; the period is the length of time required for it to complete one cycle of its variability. The value of this observation lies in the fact that periods may be measured with ease for all Cepheids, whatever their distances may be; it is their apparent brightnesses that are distance dependent. If the period of the Cepheid can be measured, then the period-luminosity law can be used to find the intrinsic brightness, i.e. the brightness the star would have if it were viewed from a standard, known distance. The brightness measured at the telescope, on the other hand, is likely to be different, since distance will be a determining factor. (It will only be the same if, by chance, it is at the exact chosen standard distance from the observer.) Hence the distance of the Cepheid may be calculated from the inverse square law of brightness. Obviously any interstellar fog will make the star appear fainter than it would be if distance were the only controlling factor, but this can generally be allowed for by making suitable

BACKGROUND TO THE BIG BANG THEORY

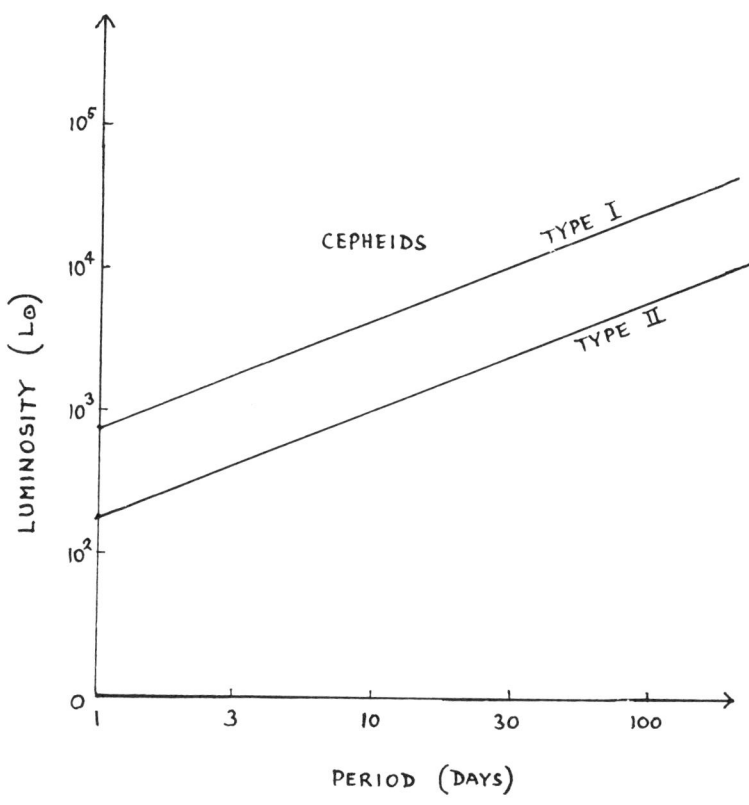

Figure 2. The Period – Luminosity Law for Cepheid Variables.

spectroscopic observations to measure the absorbing effect of such interstellar material.

In 1917 another American astronomer, Harlow Shapley, was involved in a survey of the distributions of different types of star cluster, and the period-luminosity law became one of his most useful tools. Some of the most beautiful objects in the sky are the Globular Clusters. These have star populations well in excess of one hundred thousand each and their simple symmetry makes them most attractive. It is this particular group of clusters that attracted attention because of their odd concentration in one part of the sky. It had already been noticed that almost 60 per cent of

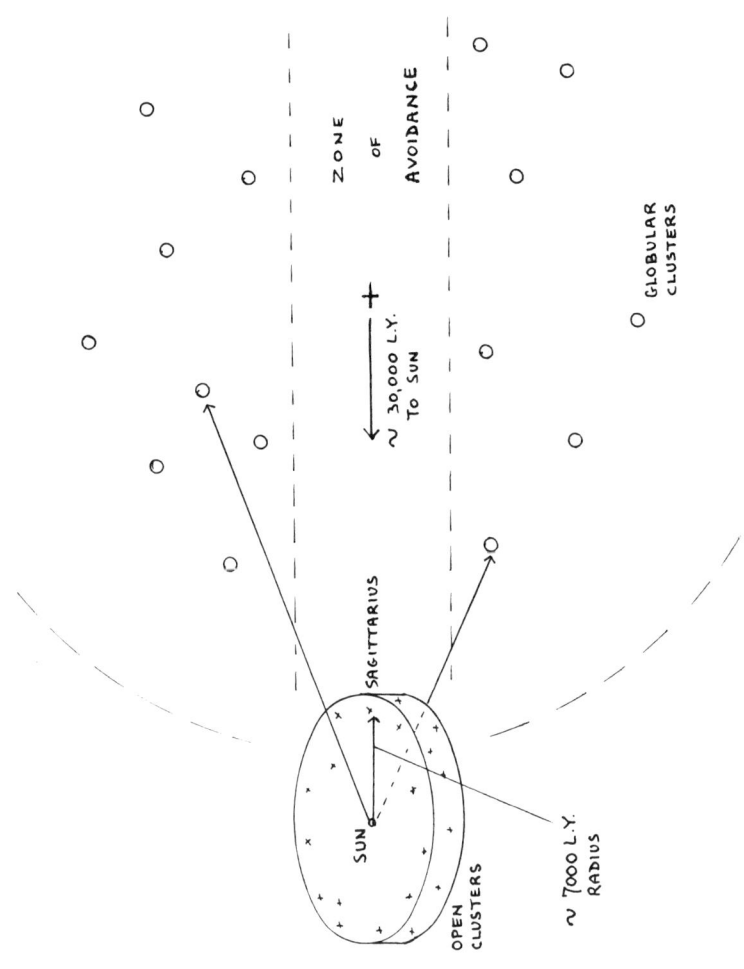

BACKGROUND TO THE BIG BANG THEORY

all the known globular clusters are to be found in the direction of the constellation Sagittarius. It seemed peculiar, however, that there should also be a zone of avoidance for these objects – none being found close to the centre line of the Sagittarius section of the Milky Way.

The situation was clarified as soon as distances were measured. Whereas the most distant objects seen in the direction of the Milky Way are about 7000 light-years away, globular clusters are, on average, more than twice this distance. If they do exist in the plane of the Galaxy, they are beyond the foreground clouds of gas and dust. It appears that the globular clusters are concentrated around a gravitational centre that we cannot see. Our Galaxy does have a central nucleus, but it is far away and hidden from our direct view.

Subsequently, Edwin Hubble identified Cepheid variables in several of the spiral nebulæ. It proved extremely difficult to photograph such faint stars, but eventually, using the 100-inch aperture Mount Wilson reflector, he succeeded in measuring their distances. The astonishing fact to emerge was that the nearest spiral, the great spiral catalogued as Messier No. 31 in Andromeda, was some 750,000 light-years from us and was therefore certainly a separate galaxy of stars having no connection whatever with our own Milky Way system. It has since been realized that Hubble had in fact used the wrong type of variable in his measurement and, after correction, the distance of M31 is now known to be 2.2 million light-years. Further analysis of star motions in the thirties strengthened the view that the Milky Way is just one galaxy out of countless millions, in which the Sun is a rather average example of a star in a family of about a hundred thousand million other stars. The opinion that the Earth has some special place in the Universe received a severe knock when it was shown that we are situated near the edge of a gigantic spiral system lost in a morass of typicality. Furthermore, our Galaxy seems to be nothing much to shout about – there are many bigger and more spectacular and, of course, some smaller. In fact it is conservatively estimated that there are probably many more galaxies within the range of our most powerful telescopes than there are stars in an individual galaxy!

The 'redshift' breakthrough

Having reached this point in our survey, we are in a position to judge the size of the problem we set ourselves when beginning to think about the origin of the Universe. We can also easily see that we are no nearer an answer than we were when the total extent of the Universe was not known.

Edwin Hubble, however, was about to discover a relationship that alters our whole view of the problem. In his study of what were then known as the 'extra-galactic nebulæ' he had used every possible observational trick to measure their distances. The results were consistent and allowed a survey to be made over hundreds of millions of light years. He had also carefully analysed the light from these galaxies, using spectroscopic techniques, and had discovered that the familiar dark lines in their spectra appear shifted towards the red, i.e. their wavelengths are lengthened in a way that depends on the distances of the galaxies from us. Thus, the further away the galaxy is, the greater is the 'redshift' in its spectral lines – a simple relationship that became known as Hubble's Law.

This law applies to all the known galaxies with the exception of the members of the local group – co-members, with the Milky Way, of our local cluster of galaxies. Messier 31, for example, shows a small blue shift, but this very exception helps us to arrive at a satisfactory interpretation of the effect.

What, then, causes the redshift?

Much debate has centred on this question, and all the possible known explanations have been tried. They range from the presence of large gravitational fields to the effects of the passage of time – the gradual reddening or decaying of light that might occur during millions of years of travel time. Such effects are not observed nor are they expected from theory.

The only successful candidate recognized at the moment is the Döppler Effect. Here the change of wavelength is caused by the relative velocity of the source of light and the observer. We are all familiar with acoustic examples of this effect, and we know that it occurs equally well for all other forms of radiation.

An example is the apparent changing pitch in the sound of an aircraft engine as it approaches, flies over and then moves away into the distance – indeed we would be very alarmed to hear the same change in pitch if we were passengers on board. Shortening of apparent wavelength indicates relative approach; lengthening indicates recession. At optical wavelengths relative approach leads to blue shift, relative recession to red shift.

BACKGROUND TO THE BIG BANG THEORY

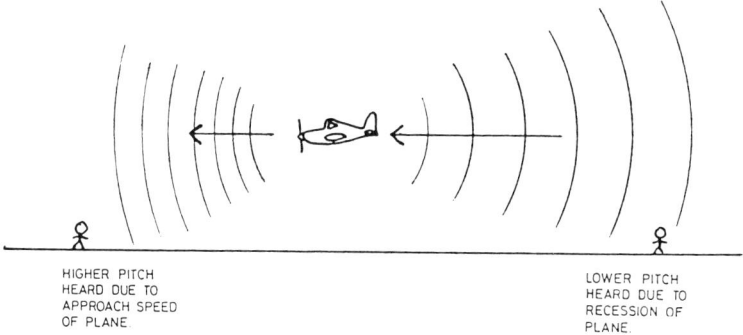

HIGHER PITCH
HEARD DUE TO
APPROACH SPEED
OF PLANE.

LOWER PITCH
HEARD DUE TO
RECESSION OF
PLANE.

Figure 4. The Doppler Effect.

If this is correct, Hubble's Law seems to show that the galaxies appear to be moving away from us and, strangely, that the further away they are; the faster they seem to be moving.

The picture that emerges is one in which we do seem to hold some sort of central position in the Universe. We look in all different directions around us and find galaxies rushing away, the more distant moving fastest. It looks as if, in some way, we are unpopular, and the rest of the Universe is trying to get away from us as fast as it can. But deeper thought about the situation shows that this interpretation is not correct.

The expanding Universe

A good way of visualizing what is happening is to use a model. Imagine a block of metal made from a regular three-dimensional crystal lattice so that we can see an even array of atoms throughout the block. Imagine we are sitting on one of the atoms somewhere near the middle of the block when someone comes along with a Bunsen burner and begins to heat the metal.

We known from elementary physics that the block will expand as it is heated; i.e. each of its dimensions will increase by a small amount. A one-inch long piece of metal may expand by one hundredth of an inch, but a two-inch block will contain twice the amount of expanding material so it will expand by two hundredths of an inch, and so on.

As we look at our nearest neighbour atom, it will move away

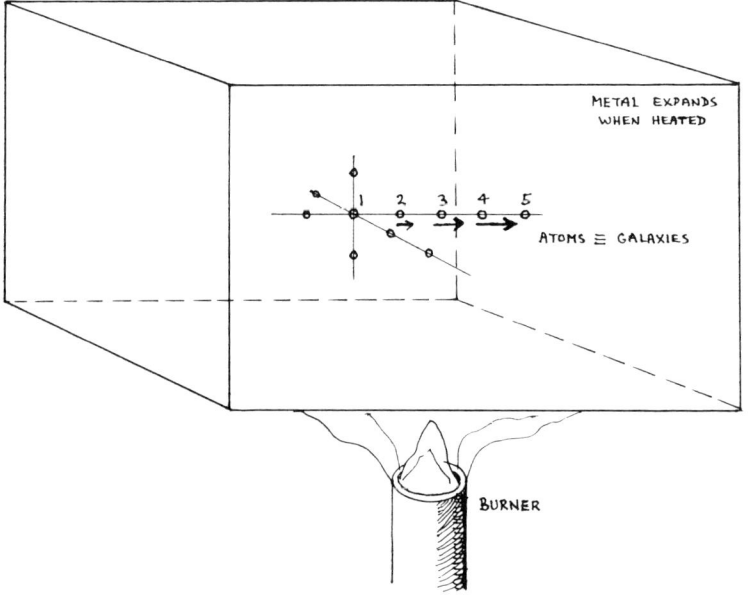

Figure 5. Thermal expansion of a block of metal as a model of the Expanding Universe. During heating the block expands in such a way that atom 2 moves away from atom 1 by a short distance. Atom 3 is twice as far away from atom 1 and therefore expands by twice the distance when heated. The heating time is constant so the apparent speed of recession of atom 3 as seen from atom 1 must be twice as much as for atom 2. This effect is the same in all directions and the same observations would be seen from whichever atom you choose as starting point. The observed speed of recession is proportional to distance from the observer.

during the heating process by a small amount. The next atom beyond this one will move twice the distance during the heating, the one beyond that three times the distance, and so on. Since the heating process takes the same time for each atom, we get the impression that the atoms expand away from us with speeds that depend on how far away they are. The more remote ones move greater distances in the same time and must therefore move faster.

This is, of course, exactly what we see for the galaxies and it means that, like the block of metal, the Universe must be expanding.

There are one or two interesting facets of this behaviour. First, we would see the same view of the surrounding atoms and their behaviour *whichever* atom we were sitting on. Second, we would only know whether we were near the centre of the block if we could see all its edges.

When we look to see if we can find an edge to the Universe we find nothing. The steady, inexorable, expansion seems to impose a limit of its own beyond which we cannot see. The scale of distance is so big that the Hubble redshift reaches truly enormous values which correspond to speeds of recession that are substantial fractions of the speed of light. What started its journey towards us as visible light waves is reddened into radio waves, and galaxies become progressively more difficult to observe as the redshift increases. If there is an edge to the Universe then it would seem to lie beyond our observational capacity. The expansion itself seems to generate an artificial boundary to what we might call our Observable Universe, but even though we may be able to see only a small fraction of the Universe within this boundary we may still be able to probe its early behaviour.

If we assume that the rate of expansion has remained constant for some time we can, as it were, work this expansion backwards as a contraction until the galaxies appear much closer together than they are now. We may even be able to go most of the way back to the time when the galaxies were all together, on top of one another, at the time of the Big Bang. Although our assumption about the constancy of the rate of expansion may be grossly incorrect we can at least demonstrate that there most probably was a Big Bang. Present calculations indicate that this was most likely some eighteen thousand million years ago, and recent observational work tends to confirm this interpretation.

Added to this is one most interesting property of the very large distance scale involved. We are dealing with information that is propagated at the speed of light, approximately one hundred and eighty six thousand miles per second. If our understanding of the distance scale is correct then we must visualize a beam of light travelling towards us for perhaps several thousand million years at this incredible speed. The message it carries refers to that galaxy as it was when the light set out towards us. In other words we see that galaxy as it was several thousand million years ago. This being the case, we should be able to look back in time simply by looking to greater distance from us.

The obvious conclusion is that if the Universe has been expanding for a long time; the galaxies further into the past should be closer together than they are now. We would expect to see an increase in the population density of galaxies the further away we look.

The underlying theory predicts an observable increase in density within the range of our most powerful radio telescopes – and this is actually confirmed by measurement. There is a significant difference between our local part of the Universe – here and now – : and the galaxies we see at great distance – there and then. We actually do have some direct evidence for an evolving Universe.

It is also highly relevant that there appear to be differences between the types of objects we find at different distances. The 'quasars', a group of objects that behave like immensely powerful nuclei of galaxies but that lack any of the other typical features of galaxies, are found apparently only at very great distance. This suggests that maybe the quasars represent an early stage in the evolution of normal galaxies. Much discussion is still taking place concerning the nature and significance of these odd objects, and the picture is far from clear.

Conclusion

Current observational research is attempting to measure exactly how the expanding Universe works. It is of great importance to know whether or not the expansion is presently slowing down. Does the Universe have sufficient mass to hold itself together in spite of the violent expansion or are some galaxies moving away at what we might call 'escape velocity' for the Universe?

If the former case applies then the expansion might eventually stop and become a contraction – with all the attendant consequences that would have. It is most unlikely that we will have the answers to questions such as these within the present century.

This then is the background to the Big Bang Theory.

In 1965 a fundamentally new piece of observational evidence, referred to as the 'three-degree K background radiation', was obtained that adds weight to our views on the Big Bang Theory. This is dealt with more fully by Alan Wright in the next section of this *Yearbook*.

As a result of careful observation, modern cosmologists can say with some confidence that the Universe seems to have originated from a single event as recently as about eighteen thousand million years ago.

What happened before that is another problem!

ALAN E. WRIGHT

A Glow from the Past

If you were to ask what was the most important cosmological discovery of the last half-century, most astronomers would probably agree that it was the detection of a faint radio 'glow' coming from the edge of the Universe. More properly, the glow is known as the 'uniform microwave sky background radiation'.

In 1964 Arno Penzias and Robert Wilson were working at the Bell telephone laboratories near Princetown, New Jersey in the USA. They were trying to make absolute strength measurements of radio sources at a wavelength of 7.3 cm, equal to a frequency of 4080 MHz. However, their equipment was plagued by excess electrical 'noise' that they couldn't explain. They worked hard to eliminate most instrumental causes of the noise, but still the problem remained. After discussing the matter with colleagues, it was realized that an explanation might be found in remnant radio radiation from the Big Bang, as predicted by some cosmologists.

Measurements since 1964 have shown that this explanation is almost certainly correct. We now know that the radiation has a distribution with frequency in the radio and far infrared typical of a perfectly emitting, or 'black', body. Confirming data comes from optical observations of molecular lines and from X-ray and gamma-ray telescopes. The black body temperature is currently determined to be 2.7 degrees Kelvin with an uncertainty of about 0.1 degrees. It is referred to loosely as the '3-degree K background radiation'.

To understand why the background radiation took so long to discover, we must understand how radio telescopes detect weak radio sources.

At radio frequencies the sky is hot. This is particularly true of very low frequencies – where our own galaxy radiates strongly, and of very high frequencies – where the radiation comes mainly from water vapour and clouds. But it is also true at the intermediate frequencies used by Penzias and Wilson for their measurements.

Even more annoying to the radio astronomer is the fact that all radio-receiver systems produce unwanted noise, arising from the random motion of electrons in the components of the receiving electronics. This noise is measured by a noise 'temperature' which is the temperature a black body would have to have to produce as much noise as the receiver. The total spurious noise temperature of a radio-telescope system – known as the 'system temperature' – is found by adding the sky-noise temperature and the receiver-system noise temperature. In a good telescope the total system noise will have a temperature of around 50 degrees Kelvin.

This noise is not so much of a problem if one is studying radio sources of small angular size, such as quasars or radio galaxies. In these cases the emission from the radio source alone can be determined by alternately directing the telescope first at the source and secondly at an off-source position just a short angular distance away in the sky. The strength of the radio source can then be found by subtracting the off-source emission from the total emission measured at the on-source position. It is a tribute to the excellence of radio astronomical systems that this method can work well even when the radio source is only one thousandth of the 'brightness' of the noise.

But this 'on-off' technique can not be used for sources that cover a large fraction of the sky, since it then becomes impossible to find a nearby, off-source position. The problem that faced Penzias and Wilson was to determine accurately the reality and strength of the faint, extended cosmic background radiation in the presence of the much-stronger system noise. Their main worry was that they had not made sufficient allowance for all the ordinary causes of this noise.

Why should we see the cosmic background radiation at all? This follows from the fact that as we look out into the Universe we are looking back in time.

During the first million years of the history of the Universe, matter existed as protons and electrons – the basic constituents of hydrogen atoms. At high temperatures – such as prevailed when the Universe was very young – hydrogen rapidly combines to form helium (and other heavier elements). But, at the present time, we find that only a small amount of matter in the Universe is in the form of helium. Why? In order to prevent the conversion of most of the hydrogen to helium, astronomers argue that there must have also been an intense field of very hot radiation present at the same time as the matter and interacting with it. This radiation would be sufficient to blast apart any heavy element atoms that tried to form.

As the Universe expanded, the radiation cooled. When its average temperature was just a few thousand degrees, the electrons and protons were able to combine into stable hydrogen atoms. At this time – which astronomers call the 'decoupling time' – the strong interaction between the matter and the radiation ceased. When we observe the cosmic microwave background, we are seeing radiation as it was at the decoupling time with a temperature of about 3000 degrees Kelvin but after being heavily redshifted.

The flux of radiation emitted by a perfectly radiating body has a distinctive form. Mathematically it is described by Planck's law. It is a normal consequence of the laws of physics that when radiation from a black body is redshifted it preserves its black body character, although at a lower temperature. The Universe has expanded since the decoupling time by a factor of about 1000, producing a redshifted temperature for the radiation of about 3 degrees Kelvin.

The redshift of about 1000 inferred for the cosmic microwave background makes it easily the most distant – and thus oldest – radio emission detected to date. In comparison the redshifts of even the highest redshift quasars (around 4) are insignificant.

It is one of the curiosities of astronomy, however, that this highly redshifted sky background radiation provides information about such a local phenomenon as the absolute motion of the Earth through space.

The special (or restricted) theory of relativity states that all motion is relative, depending on the space and time frames used by the observer. But the special theory is only valid for small-scale phenomena. On the cosmic scale, space and time are absolute and it is meaningful, and necessary, to talk about absolute motion relative to the 'fabric' of the Universe. Any absolute motion the Earth has relative to this fabric – or, more accurately, relative to a 'co-moving' observer – will cause the redshift, and hence the temperature we measure for the background radiation, to be different in different directions. This temperature will be highest for parts of the sky lying in the direction of the Earth's absolute motion and least in the opposite direction.

But to determine the Earth's motion requires measurements of the highest precision. For example, a speed of 300 kilometres per second is only 0.1 per cent of the velocity of light. This means that the microwave background temperature will be changed also by only 0.1 per cent; that is, about 0.003 degrees. At the present, the best measurements indicate that the Earth is moving at a speed of about 600 kilometres per second towards a point in the direction of the constellation of Hydra. Part of this motion (about 200 kilometres per second) is, of course, due to the Earth and Sun's motion around our Galaxy. However, it is also clear that our Galaxy is moving as well relative to the absolute frame of the background radiation.

Perhaps even more interesting than the slight variations in the temperature of the background radiation with direction is the background's overall uniformity, or 'isotropy'. We observe essentially identical temperatures in widely different parts of the Universe. So widely separated, in fact, that they can never have communicated with each other, even by radiation or particles travelling at the speed of light. It is as though we gave a fancy dress party and everybody turned up in the same costume! It could, of course, be due to chance. But we would strongly suspect that our guests had somehow decided – or been contacted and told – to dress similarly. In the cosmological case, all our understanding clearly leads us to believe that the widely separated parts of the Universe can *NEVER* have been in contact.

Why then are they so similar? Astronomers can offer no explanation – at least on the basis of models of the Universe which expand in a regular manner. Some cosmologists have discussed Universe models that have a rapid burst of expansion around the decoupling time – the so-called 'inflationary' models. These models do alleviate some of the problems of the isotropy of the background radiation but they remain highly speculative.

What *IS* clear is that the faint hiss of noise first heard by Penzias and Wilson in their New Jersey radio-telescope was as important a discovery as Hubble's realization that the Universe expands. It appears to have confirmed beyond doubt that the Universe has not existed for ever, as per the Steady State theories but, instead, began with a much more spectacular Big Bang.

PATRICK MOORE

Supernova!

A supernova explosion is the most violent outburst known to us. Unlike an ordinary nova, it involves the death of a star – either a white dwarf which blows itself to pieces, or a more massive star which has come to the end of its main career and destroys itself, leaving only a tiny, super-massive remnant. At the peak of its outburst, a supernova shines many times more brilliantly than the Sun, so that it can be seen over vast distances; amateurs now hunt for supernovæ in external galaxies, in the manner described by Ron Arbour in the 1987 *Yearbook*. But no supernovæ in our own Galaxy have been seen since telescopes were invented, and astronomers would dearly like to be able to study one from comparatively close range.

This has still not happened, but we have at least been privileged to watch the next best thing: a supernova in the Large Magellanic Cloud (LMC for short). It was discovered in February 1987, and took astronomers completely by surprise.

The LMC is much the brightest of the outer galaxies, because it is relatively nearby. Its distance is usually given as about 170,000 light-years, though astronomers at the Royal Greenwich Observatory prefer the rather lower value of 155,000 light-years. This is not much compared with, say, the 2,200,000 light-years of the Andromeda Spiral, and it is not surprising that the LMC is a prominent naked-eye object. Superficially it looks rather like a broken-off portion of the Milky Way, and even moonlight will not drown it. It lies in the far south of the sky, fairly close to the southern celestial pole, so that it is never visible from the latitudes of Europe or the United States; even from Hawaii it merely skirts the horizon – a fact which is widely regretted in the Old World. On the other hand it is favourably placed from countries such as South Africa, Australia, and New Zealand, and is one of the main features of the southern night sky. It is, of course, a member of our Local Group of galaxies; so is the smaller and slightly more distant Small Magellanic Cloud (SMC). Incidentally, the clouds are named in honour of the Portuguese round-the-globe explorer Magellan, who certainly observed them. An older name for them was the Nubeculæ.

The Clouds are important because they contain objects of all kinds: giant and dwarf stars, clusters, nebulæ, novæ – in fact everything which we see in our own Galaxy. It was by studying short-period variables in the SMC that Henrietta Leavitt laid the foundations of the Cepheid period-luminosity law which has been of such immense value in modern astrophysics; the LMC contains the largest known gaseous nebula, the Tarantula, and also the exceptionally luminous star S Doradûs, which is at least a million times brighter than the Sun (though admittedly even S Doradûs cannot rival a member of our own Galaxy, the unique Eta Carinæ). There are also fairly frequent novæ in the LMC, but until February of this year no supernova had been observed there.

Supernovæ are of two main types. With Type I we have a binary system, made up of a normal star together with a white dwarf. The white dwarf pulls material away from its larger, less dense companion, and eventually accumulates so much material that a nuclear outburst takes place – and the dwarf is literaly blown to pieces. With Type II we have only one star, which runs out of 'fuel' and implodes, so that after a tremendous shock-wave most of the star's material is dissipated in space, leaving a remnant in the form of either a neutron star or, possibly, a black hole. But our knowledge has always been limited by the fact that we have had to content ourselves with watching supernova outbursts many millions of light-years away.

During the past thousand years four supernovæ have been definitely observed in our Galaxy. One, in 1006, flared up in the southern constellation of Lupus, and seems to have become as bright as the quarter-moon, but unfortunately it is poorly documented. Next came the star of 1054, which has left the remnant which we know today as the Crab Nebula – a patch of expanding gas in the midst of which is a neutron star or pulsar, acting as the Crab's 'power-house'. The distance is some 6,000 light-years, and the value of the Crab in modern research can hardly be over-estimated, if only because it sends out radiations at almost all wavelengths ranging from radio waves to X-rays. Then, in 1572, came the supernova we now know as B Cassiopeiæ; it was carefully watched by the great Danish astronomer Tycho Brahe, and became brilliant enough to be seen with the naked eye in broad daylight. It has left no pulsar, though we can see faint wisps of nebulosity and can also pick up radio emissions. This is also true of Kepler's star of 1604, in Ophiuchus, which was the last supernova which we can prove to have been visible with the naked eye. Sadly, the telescope did not come into use until some years later, and since then supernovæ in our own Galaxy have been lacking.

There was, however, an outburst in the Andromeda Spiral, M.31, in 1885, which has been listed as S Andromedæ. It reached the sixth magnitude, and was therefore on the fringe of naked-eye visibility, but at the time its nature was not appreciated; it was even thought likely that the Spiral itself was a feature of our Galaxy rather than an external system, and the chance was missed. (Incidentally, one of the independent discoverers was a Hungarian baroness who was holding a house-party and had set up a small telescope on the lawn of her castle!) We now think that S Andromedæ was atypical, and decidedly underluminous by supernova standards, though all trace of it has long since been lost and we will never know much more about it.

All that astronomers could do was to wait and hope. The long wait ended on February 24, 1987, with the outburst in the LMC. There were three independent discoverers: Ian Shelton and Oscar Duhalde, at the Las Campanas Observatory in Chile, and the veteran New Zealand amateur variable star observer Albert Jones. Robert McNaught, at the Siding Spring Observatory in Australia, actually photographed it earlier, but failed to develop his plate promptly, and so missed the opportunity of being sole discoverer.

When first seen, the supernova – officially designated SN 1987A – was of around the fourth magnitude, and so was clearly visible without optical aid, making it much the most conspicuous super-

Figure 1 (opposite). This photo of the supernova was taken by the author on May 21, 1987 from South Africa using an unguided camera. The ASA 1600 film was exposed for 40 seconds. It shows the Large Cloud of Magellan as well as the supernova. This is a good example of what can be achieved with an ordinary camera.

nova since Kepler's Star of well over three centuries earlier. Needless to say, all southern-hemisphere observatories concentrated upon it, and waited to see what would happen, in particular, was it of Type I or Type II? And would it be possible to track down the progenitor star?

The light-curves of Types I and II supernovæ differ markedly, but SN 1987A did not seem to fit happily into either category. At first it was thought that it must be of Type II, and there were suggestions that the progenitor star was a blue supergiant which had been catalogued by Nicholas Sanduleak in 1969, and was known as Sanduleak-69°202. After a few days observations from the veteran satellite IUE (International Ultra-violet Explorer) indicated that this could not be so, but after a further delay it became clear that the IUE work had been wrongly interpreted, and the suspect star became a favoured candidate again. For some time the magnitude hovered around 4, which indicated that, as with S Andromedæ, the outburst was underluminous for a supernova; yet it could not be a normal nova, which would have remained well below naked-eye visibility. By mid-May the magnitude had risen to above 3, and astronomers were frankly puzzled.

Of special importance were two bursts of neutrinos which came at different times. A supernova explosion would be expected to produce showers of these strange chargeless, virtually massless particles, and it was suggested that the two records could indicate different stages in the supernova's development – the first when the core of the progenitor collapsed to become a neutron star, and the second when the neutron star collapsed further to produce a black hole. But it was all decidedly uncertain, and there was always the chance that we were seeing something quite unfamiliar – a Type III supernova, if you like.

Within a few months, possibly before this edition of the *Yearbook* appears in print we should know much more. Everything will depend upon how the outburst progresses, and what its spectra shows us. There is the exciting possibility that we may really have witnessed the birth of a black hole, in which case the neutrino bursts will allow us to pinpoint the exact moment of its formation. But at least it is fair to say that SN 1987A is one of the most exciting astronomical events for many years, and is providing us with unrivalled opportunities for research. Most astronomers will admit, perhaps rather grudgingly, that it is far more important than the return of Halley's Comet!

PART THREE

Miscellaneous

Some Interesting Telescopic Variable Stars

Star	R.A. h	m	Dec. °		mag. range	Period days	Remarks
R Andromedæ	0	22	+38	18	6.1–14.9	409	
W Andromedæ	2	14	+44	4	6.7–14.5	397	
Theta Apodis	14	00	−76	33	6.4– 8.6	119	Semi-regular.
R Aquilæ	19	4	+ 8	9	5.7–12.0	300	
R Arietis	2	13	+24	50	7.5–13.7	189	
R Aræ	16	35	−56	54	5.9– 6.9	4	Algol type.
R Aurigæ	5	13	+53	32	6.7–13.7	459	
R Boötis	14	35	+26	57	6.7–12.8	223	
Eta Carinæ	10	43	−59	25	−0.8– 7.9	–	Unique erratic variable.
I Carinæ	09	43	−62	34	3.9–10.0	381	
R Cassiopeiæ	23	56	+51	6	5.5–13.0	431	
T Cassiopeiæ	0	20	+55	31	7.3–12.4	445	
X Centauri	11	46	−41	28	7.0–13.9	315	
T Centauri	13	38	−33	21	5.5– 9.0	91	Semi-regular.
T Cephei	21	9	+68	17	5.4–11.0	390	
R Crucis	12	20	−61	21	6.9– 8.0	5	Cepheid.
Omicron Ceti	2	17	− 3	12	2.0–10.1	331	Mira.
R Coronæ Borealis	15	46	+28	18	5.8–14.8	–	Irregular
W Coronæ Borealis	16	16	+37	55	7.8–14.3	238	
R Cygni	19	35	+50	5	6.5–14.2	426	
U Cygni	20	18	+47	44	6.7–11.4	465	
W Cygni	21	34	+45	9	5.0– 7.6	131	
S S Cygni	21	41	+43	21	8.2–12.1	–	Irregular.
Chi Cygni	19	49	+32	47	3.3–14.2	407	Near Eta.
Beta Doradûs	05	33	−62	31	4.5– 5.7	9	Cepheid.
R Draconis	16	32	+66	52	6.9–13.0	246	
R Geminorum	7	4	+22	47	6.0–14.0	370	
U Geminorum	7	52	+22	8	8.8–14.4	–	Irregular.
R Gruis	21	45	−47	09	7.4–14.9	333	
S Gruis	22	23	−48	41	6.0–15.0	410	
S Herculis	16	50	+15	2	7.0–13.8	307	
U Herculis	16	23	+19	0	7.0–13.4	406	
R Hydræ	13	27	−23	1	4.0–10.0	386	
R Leonis	9	45	+11	40	5.4–10.5	313	Near 18, 19.
X Leonis	9	48	+12	7	12.0–15.1	–	Irregular. (U Gem type)
R Leporis	4	57	−14	53	5.9–10.5	432	'Crimson star.'
R Lyncis	6	57	+55	24	7.2–14.0	379	
W Lyræ	18	13	+36	39	7.9–13.0	196	

183

Star	R.A. h	R.A. m	Dec. °	Dec.	mag. range	Period days	Remarks
T Normæ	15	40	−54	50	6.2−13.4	293	
H R Delphini	20	40	+18	58	3.6− ?	−	Nova, 1967.
S Octantis	17	46	−85	48	7.4−14.0	259	
U Orionis	5	53	+20	10	5.3−12.6	372	
Kappa Pavonis	18	51	−67	18	4.0− 5.5	9	W Virginus
R Pegasi	23	4	+10	16	7.1−13.8	378	
S Persei	2	19	+58	22	7.9−11.1	810	Semi-regular.
R Sculptoris	01	24	−32	48	5.8− 7.7	363	Semi-regular.
R Phœnicis	23	53	−50	05	7.5−14.4	268	
Zeta Phœnicis	01	06	−55	31	3.6− 4.1	1	Algol type.
R Pictoris	04	44	−49	20	6.7−10.0	171	Semi-regular.
L² Puppis	07	12	−44	33	2.6− 6.0	141	Semi-regular.
Z Puppis	07	30	−20	33	7.2−14.6	510	
T Pyxidis	09	02	−32	11	7.0−14.0	−	Recurrent nova (1920, 1944)
R Scuti	18	45	− 5	46	5.0− 8.4	144	
R Serpentis	15	48	+15	17	5.7−14.4	357	
S U Tauri	5	46	+19	3	9.2−16.0	−	Irregular. (R CrB type).
R Ursæ Majoris	10	41	+69	2	6.7−13.4	302	
S Ursæ Majoris	12	42	+61	22	7.4−12.3	226	
T Ursæ Majoris	12	34	+59	46	6.6−13.4	257	
S Virginis	13	30	− 6	56	6.3−13.2	380	
R Vulpeculæ	21	2	+23	38	8.1−12.6	137	

Note: Unless otherwise stated, all these variables are of the Mira type.

Some Interesting Double Stars

We are very grateful to Robert Argyle for this revised list of double stars, which is up to date.

Name	Magnitudes	Separation "	Position angle °	Remarks
Gamma Andromedæ	3.0, 5.0	9.4	064	Yellow, blue. B is again double (0″.5) but needs larger telescope.
Zeta Aquarii	4.4, 4.6	1.8	217	Becoming more difficult.
Gamma Arietis	4.2, 4.4	7.8	000	Very easy.
Theta Aurigæ	2.7, 7.2	3.5	313	Stiff test for 3″0G.
Delta Boötis	3.2, 7.4	105	079	Fixed.
Epsilon Boötis	3.0, 6.3	2.8	335	Yellow, blue. Fine pair.
Kappa Boötis	5.1, 7.2	13.6	237	Easy.
Zeta Cancri	5.6, 6.1	5.6	085	Again double.
Iota Cancri	4.4, 6.5	31	307	Easy. Yellow, blue.
Alpha Canum Ven.	3.2, 5.7	19.6	228	Easy. Yellowish, bluish.
Alpha Capricorni	3.3, 4.2	376	291	Naked-eye pair.
Eta Cassiopeiæ	3.7, 7.4	12.2	310	Easy. Creamy, bluish.
Beta Cephei	3.3, 8.0	14	250	Easy with a 3 in.
Delta Cephei	var, 7.5	41	192	Very easy.
Alpha Centauri	0.0, 1.7	21.7	212	Very easy. Binary, period 80 years.
Xi Cephei	4.7, 6.5	6.3	270	Reasonably easy.
Gamma Ceti	3.7, 6.2	2.9	294	Not too easy.
Alpha Circini	3.4, 8.8	15.7	230	PA slowly decreasing.
Zeta Coronæ Bor.	4.0, 4.9	6.3	305	PA slowly increasing.
Delta Corvi	3.0, 8.5	24	214	Easy with 3 in.
Alpha Crucis	1.6, 2.1	4.7	114	Third star in low-power field.
Gamma Crucis	1.6, 6.7	111	212	Wide optical pair.
Beta Cygni	3.0, 5.3	34.3	055	Glorious. Yellow, blue.
61 Cygni	5.3, 5.9	29	147	Slowly widening. (Add .5)
Gamma Delphini	4.0, 5.0	9.6	268	Easy. Yellow, greenish.
Nu Draconis	4.6, 4.6	62	312	Naked-eye pair.
Alpha Geminorum	2.0, 2.8	2.6	085	Becoming easier.
Delta Geminorum	3.2, 8.2	6.5	120	Not too easy.

Name	Magnitudes	Separation "	Position angle °	Remarks
Alpha Herculis	var, 6.1	4.6	106	Red, green.
Delta Herculis	3.0, 7.5	8.6	262	Optical pair.
Zeta Herculis	3.0, 6.5	1.5	110	Fine, rapid binary (34y)
Gamma Leonis	2.6, 3.8	4.4	123	Binary; 619 years.
Alpha Lyræ	0.0, 10.5	73	180	Optical. B is faint.
Epsilon Lyræ	4.6, 6.3	2.6	356	Quadruple. Both pairs.
	4.9, 5.2	2.2	093	separable with 3 in.
Zeta Lyræ	4.2, 5.5	44	149	Fixed. Easy double.
Beta Orionis	0.1, 6.7	9.5	205	Can be split with 3 in.
Iota Orionis	3.2, 7.3	11.8	141	Enmeshed in nebulosity.
Theta Orionis	6.8, 7.9	8.7	032	Trapezium in M. 42.
	6.8, 5.4	13.4	241	
Sigma Orionis	4.0, 10.3	11.1	236	Quadruple. C is rather
	6.8, 8.0	30.1	231	faint in small apertures.
Zeta Orionis	2.0, 4.2	2.4	162	Can be split with 3 in.
Eta Persei	4.0, 8.5	28.5	300	Yellow, bluish.
Beta Phœnicis	4.1, 4.1	1.1	319	Slowly closing.
Beta Piscis Austr.	4.4, 7.9	30.4	172	Optical pair. Fixed.
Alpha Piscium	4.3, 5.3	1.9	283	Binary; 720 years.
Kappa Puppis	4.5, 4.6	9.8	318	Again double.
Alpha Scorpii	0.9, 6.8	3.0	275	Red, green.
Nu Scorpii	4.2, 6.5	42	336	Both again double.
Theta Serpentis	4.1, 4.1	22.3	103	Very easy.
Alpha Tauri	0.8, 11.2	131	032	Wide, but B very faint in small telescopes.
Beta Tucanæ	4.5, 4.5	27.1	170	Both again double.
Zeta Ursæ Majoris	2.1, 4.2	14.4	151	Very easy. Naked-eye pair with Alcor.
Alpha Ursæ Minoris	2.0, 9.0	18.3	217	Can be seen with 3 in.
Gamma Virginis	3.6, 3.7	3.5	292	Binary; 171 years. Closing.
Theta Virginis	4.0, 9.0	7.1	343	Not too easy.
Gamma Volantis	3.9, 5.8	13.8	299	Very slow binary.

Some Interesting Nebulæ and Clusters

Object	R.A.		Dec.		Remarks
	h	m	°	'	
M.31 Andromedæ	00	40.7	+41	05	Great Galaxy, visible to naked eye.
H.VIII 78 Cassiopeiæ	00	41.3	+61	36	Fine cluster, between Gamma and Kappa Cassiopeiæ.
M.33 Trianguli	01	31.8	+30	28	Spiral. Difficult with small apertures.
H.VI 33 4 Persei	02	18.3	+56	59	Double cluster; Sword-handle.
△142 Doradûs	05	39.1	−69	09	Looped nebula round 30 Doradûs. Naked-eye. In Large Cloud of Magellan.
M.1 Tauri	05	32.3	+22	00	Crab Nebula, near Zeta Tauri.
M.42 Orionis	05	33.4	−05	24	Great Nebula. Contains the famous Trapezium, Theta Orionis.
M.35 Geminorum	06	06.5	+24	21	Open cluster near Eta Geminorum.
H.VII 2 Monocerotis	06	30.7	+04	53	Open cluster, just visible to naked eye.
M.41 Canis Majoris	06	45.5	−20	42	Open cluster, just visible to naked eye.
M.47 Puppis	07	34.3	−14	22	Mag. 5,2. Loose cluster.
H.IV 64 Puppis	07	39.6	−18	05	Bright planetary in rich neighbourhood.
M.46 Puppis	07	39.5	−14	42	Open cluster.
M.44 Cancri	08	38	+20	07	Præsepe. Open cluster near Delta Cancri. Visible to naked eye.
M.97 Ursæ Majoris	11	12.6	+55	13	Owl Nebula, diameter 3'. Planetary.
Kappa Crucis	12	50.7	−60	05	"Jewel Box"; open cluster, with stars of contrasting colours.
M.3 Can. Ven.	13	40.6	+28	34	Bright globular.
Omega Centauri	13	23.7	−47	03	Finest of all globulars. Easy with naked eye.
M.80 Scorpii	16	14.9	−22	53	Globular, between Antares and Beta Scorpionis.
M.4 Scorpii	16	21.5	−26	26	Open cluster close to Antares.
M.13 Herculis	16	40	+36	31	Globular. Just visible to naked eye.
M.92 Herculis	17	16.1	+43	11	Globular. Between Iota and Eta Herculis.
M.6 Scorpii	17	36.8	−32	11	Open cluster; naked-eye.
M.7 Scorpii	17	50.6	−34	48	Very bright open cluster; naked eye.
M.23 Sagittarii	17	54.8	−19	01	Open cluster nearly 50' in diameter.
H.IV 37 Draconis	17	58.6	+66	38	Bright Planetary.
M.8 Sagittarii	18	01.4	−24	23	Lagoon Nebula. Gaseous. Just visible with naked eye.
NGC 6572 Ophiuchi	18	10.9	+06	50	Bright planetary, between Beta Ophiuchi and Zeta Aquilæ.

M.17 Sagittarii	18	18.8	−16	12 Omega Nebula. Gaseous. Large and bright.
M.11 Scuti	18	49.0	−06	19 Wild Duck. Bright open cluster.
˙M.57 Lyræ	18	52.6	+32	59 Ring Nebula. Brightest of planetaries.
M.27 Vulpeculæ	19	58.1	+22	37 Dumb-bell Nebula, near Gamma Sagittæ.
H.IV 1 Aquarii	21	02.1	−11	31 Bright planetary near Nu Aquarii.
M.15 Pegasi	21	28.3	+12	01 Bright globular, near Epsilon Pegasi.
M.39 Cygni	21	31.0	+48	17 Open cluster between Deneb and Alpha Lacertæ. Well seen with low powers.

Our Contributors

Dr David A. Allen, our most regular contributor, continues his researches at Siding Spring in Australia. No issue of the *Yearbook* would be complete without him.

Professor Alec Boksenberg, FRS is the Director of the Royal Greenwich Observatory. He has developed many new astronomical instruments, and is acknowledged as the world leader in this field.

Dr Garry Hunt is author, broadcaster, scientist; now with PA Computers and Telecommunications where he is responsible for developing the company's Space consulting activities. During his career he has consulted regularly to NASA and ESA, held visiting professorships in the USA, Australia, and Egypt; until 1986 he was Director of the Centre for Remote Sensing at Imperial College of Science and Technology, University of London. He is a member of the International Academy of Astronautics, an honorary member of the Australian Space Society, a member of the Board of advisors of the Planetary Society, a member of the Space Society (UK), a Fellow of the British Interplanetary Society; President of the Commission on the Planets (16) of the International Astronomical Union. President of the Commission on Planetary Atmospheres and their Evolution of the International Association for Meteorology and Atmospheric Physics. He has been a NASA experimenter on the Viking, ERBE and Voyager missions receiving 3 NASA awards for his Voyager activities and the Gaskell Memorial Medal of the Royal Meteorological Society for his planetary studies.

Dr Ron C. Maddison, of the University of Keele, is an astrophysicist and cosmologist who is also very much concerned with astronomical education. He is very well known as a lecturer and broadcaster as well as for his technical work.

Dr Paul Murdin is a Cambridge graduate, and from the Royal Greenwich Observatory has been in charge of the British telescopes at La Palma. He has carried out fundamental researches in astrophysics, and is also a leading astronomical author of popular books as well as technical works.

Dr Alan E. Wright, a Cambridge graduate, is essentially a radio astronomer, and carries out his work at the Parkes Radio Astronomy Observatory in New South Wales. He is also a specialist in many branches of astrophysics.

Notes for Readers

There is now a permanent public exhibition at the Royal Greenwich Observatory, Herstmonceux Castle, Hailsham, East Sussex. It is open on weekdays between 2 and 5.30 p.m., and on Saturdays, Sundays and public holidays from 10.30 a.m. to 5.30 p.m. The Exhibition covers the fields of modern astronomy, the development of telescopes, the history of the Observatory and of the Castle. There is a tea room (June-September), a souvenir shop, and free parking.

The William Herschel Society maintains the museum now established at 19 New King Street, Bath – the only surviving Herschel house. It also undertakes activities of various kinds. New members would be welcome; those interested are asked to contact Dr L. Hilliard at 2 Lambridge, London Road, Bath.